ARROW

THE POWER AND POISON OF STORY

WILLIAM GADEA

First edition, September 30, 2025
ISBN: 979-8-9988935-0-6
Library of Congress Control Number:

Cover design by William Gadea and Jeronimo Adrien
Interior design by Lorna Reid - Reedsy
Illustrations by Manuela Martin
Edited by James Kingsland
Copy editing by Jennifer Griffith

Published by Marrow Publishing
New York, NY

For information, contact:
wgadea@williamgadea.com
www.williamgadea.com

CONTENTS

To my teachers.

INTRODUCTION

THE STORY ANIMAL

Imagine our hunter-gatherer ancestors circled around a fire at night, after a long day when their hunting efforts failed. A young man, barely more than a boy, shifts in his place with an energy he can't quite contain. Suddenly, he bolts to his feet, commanding the attention of his tribe.

"I speak!" says Shiki.

"You have not enough hair on your chin to speak," sneers his uncle, Murrak. The older man wins a couple of chuckles with this, but they are quickly quieted when Murrak's mother – the tribe's Matriarch– hisses her disapproval of his words. The tribe falls silent, turning their attention to Shiki.

"The deer was swift of foot today. As soon as they catch our scent, they flee. We cannot come close enough to spear them. So our hunt fails. We gather like this with no meat on the fire.

"Remember our hunt last moon, on the other side of the river?" Shiki gestures eastward. "Some of you were on one side of a bluff, and you scared a buck so that it jumped over the bluff toward me. My hunt was simple." Shiki pantomimes his awed reaction as a deer heads toward him. "He practically impaled himself on my spear!" Shiki pantomimes holding a spear vertically by his side, and having a deer jump on top of it. The tribe laughs at this, but it is laughter that reassures Shiki. They know he is clowning and exaggerating for effect – spearing that buck was not quite as easy as that.

"We helped you there," Murrak interjects, claiming his bit of credit.

Shiki pounces. "Yes, you did, uncle! You did." He pauses for effect. "So why not work together like that again?

"The deer always graze on the field, near the ravine where the creek runs. What if a pair of us waited in the ravine, hidden? Then the rest of us approach the deer in an arc, and rouse them toward the ravine?" Shiki spreads his arms to demonstrate a corralling action. Then he brings his hands together to mimic the tight walls of the ravine. "Enclosed in that narrow space, the deer would not escape. It would be an easy hunt." Shiki feels the quality of the attention changing in the group; he wants to finish strong.

"As a tribe, we always work together. That's how we drove off the valley tribe, when they tried to encroach on our hunting grounds. Every day, each of us contributes to the good of all – that's what makes us strong and special. Why not cooperate in the hunt this way? If we did, there might be a carcass over that fire there. We would have enough meat to sate us all!"

Shiki looks at the faces around the fire: the broad smile of his best friend, the bright eyes of his pretty cousin, the affirmative squint of the Matriarch, and even the acquiescent stone face of Murrak – he has won them over.

Shiki poses a problem, tells a story that indicates a possible solution, and projects actions into the future that he hopes will provide a bounty. This story might entertain, but its objective is to coordinate the action of members of his species. It is a story that seeks to influence.

Our storytelling ancestor could have offered his idea on the walk back from that day's hunt, but that likely would not have been as effective. Instead, he waits until he has his whole tribe's attention at the fireside that night. He primes them for his proposal, pointing out that they are hungry for meat. He builds his relationship with his audience, by pointing out his debt to them when they helped him spear a deer. Finally, he dramatizes a vision of future triumph. And he activates feelings of pride and tribal affiliation.

Shiki is not just offering an idea. He is creating a world of actions, of causes and effects, of emotions, of values. He is inviting the others into a mental world.

Are humans the only animals capable of such storytelling? Around that fire, there were likely many other species nearby using signals to communicate and coordinate. Birds have specific calls to warn of predators. Crickets rub their wings together to signal their availability for mating. Worker bees, after foraging, perform intricate dances to inform their hive-mates about the location of nectar. These behaviors are thought to be mostly hardwired rather than learned.

But what about the learned behaviors we call culture? Until recently, culture was considered uniquely human, but recent studies of monkeys, whales and other animals prove that there is a reservoir of learning that is passed on from older animals to younger ones by means of demonstration and imitation.[1]

And stories? What about a narrated sequence of events, like the fireside story told by our ancestor? We have no evidence of any other animal being able to perform this activity. (Which of course does not mean we will never have it; as it was with learning, claims for human uniqueness have tended to have a short half-life.) A story is not just mimicry or symbolic communication; as we shall define it, a story is a representation of a causal sequence of events. Such a representation can carry a lot more than its bare facts; it can present a whole model of life that allows us to teach, influence, and coordinate others.

Storytelling isn't just a nice-to-have feature of human biology. It is our species' super-power. It is what has allowed us to build civilizations. Since our fictional hunter-gatherer ancestor told that story around the fireside, there have been many world-shaping stories told: by chieftains to explain the benefits of loyalty, by courtiers to provoke a war-making anger, by entrepreneurs to detail the benefits of a product. Every societal re-organization, every new invention, every claim to the mantle of leadership has required a story to support it.

But stories of influence do not live just on the stage of history. When we commit a social faux pas, we provide a clean narrative to excuse our lapse. When we seek to deepen a friendship, we recount its history in celebration. When we are on a first date, we try to outline our biography in a way that is most flattering (but still avoids conceit!). To be the most effective social animals we can be, we tell tales... and some of them are even true.

That is the Power of Story. But is there also Poison in Story?

Turning Point

A spiritual crisis is a broken story.

In my life, spiritual crises have usually been preceded by romantic disappointments. I'm not quite sure whether this fact adds or detracts credibility from the substance of these crises, but I'll let you decide.

I place one such inciting incident in New York's Central Park, mid-2000s. A vibrant, brilliant woman lets me know she has chosen another. I instantly feel my self shrinking into me, like a deflating balloon wrapped around the pipe end of a vacuum cleaner. The space between me and the world feels like a hopeless chasm. I hear her trying to be kind, but not kind enough. I start to wonder why I took so much trouble deciding what to wear that day.

The woman happened to be a Christian and the man she chose was Christian. So you might think that my next move was a pathetic attempt to win her back. It was a Sunday, so I, the lifelong atheist, went to Church. Actually, I don't think I was trying to win her back – but I did consider that her religious practice was an ingredient that deserved to be sampled without prejudice – and that morning, I was definitely in need of some sort of salvation.

So it was that I went to Church. My Christian experiment lasted for maybe six weeks. My theory was that you had to begin with the relationship – that relationship needed to precede belief. So I prayed to God for belief in God, I prayed for faith. I did get other things – consolation, the warmth of community, tears at communion – but faith was not among them.

Moments like these force self-accounting, and I did plenty of it. The ledger was damning. Middle-aged man, living alone in New York City. A failed playwright who had turned to animation in search of a commercial craft, and seemed to be failing at that too. Overweight, drinking too much, and sinking into serious depression.

I tried a new, different remedy: I walked into a Zendo, a Zen Buddhist place of practice.

Even the first step over the threshold had a therapeutic effect. A calmness filled the air. The decorations were minimal. Rows of zabutons and zafus – pairings of sitting mats and cushions – lined the walls. A senior student took another visitor and me into a side room to give us a crash course in Zen Buddhism and Zazen, their form of meditation.

We sat down on our cushions. He smiled at us and paused for a moment.

"The Buddha thinks you're perfect just the way you are," he said. *Not sure I agree*, I thought, *but tell me more*.

And then he spoke of the Buddha's Four Noble Truths.

First, that life is *dukkha*, meaning unease or dissatisfaction. (I had read about Buddhism before and seen this translated as Suffering. Even a depressive can see that's not an accurate description of life. Dissatisfaction seemed closer to the mark.)

Second, that it is so because of *upadana*, meaning clinging or attachment. (I had read this translated as desire, which was an even more disastrous mistranslation. All the meditation in the world won't stop you from desiring.)

Third, if you cease the clinging of *upadana*, you can stop the pain of *dukkha*.

Fourth and finally, that it is possible to do this by following a Buddhist methodology of practice.

I was an easy sell. It was clear to me that the stories I told myself about myself were my *upadana*, or at least related to it. When I compared my actual life with my dreams – of theatrical achievement, of professional success, of romantic harbor – my life came up short. A carousel of self-criticism kept turning in my head. If I could stop the spinning and get off, if I could seize each moment of my life and discover it freshly, then there might be an opportunity for a new life.

This was a different tack than I usually tried. I had already figured out that the difference between my actual life and my inner story-life was the source of my discontent, but the answer had always been either to try to tweak the life or try to tweak the story. The idea here was to break the connection – to say "Story, you are not me."

I am not the most conscientious Zen practitioner, but since that day I have tried to sit on a cushion once a day and follow the other prescriptions of the practice.

I hesitate to advertise too loudly the benefits. The idea of Zen Buddhism is to live in the moment, so to practice with a "gaining mind" is counterproductive. When we are sitting on a cushion with an idea in our minds of what we wish meditation will grant us, then we are not in the moment, not truly on the cushion. Instead of untying a strand of *upadana*, we are weaving a new one. Therefore, to trumpet one's gains might be to encourage someone to go down a wrong path, and cheat them of their own gains.

Still, an honest accounting requires me to say: I've lived most of my life wishing for a transformation, for a magical moment when larva becomes butterfly, when a new state of being is entered. I've only lived one true transformation, and it began after I stepped into that Zendo.

In the decade after I took that step I started a successful animation company. When I became a writer-director-producer of animation, my broad but shallow knowledge of animation was finally a boon instead of a handicap. I met and married the woman who is still my wife, and every day is a (mainly joyful) struggle to truly be present with her. And I managed to shed those depressionary spirals for the mixture of contentment and common sadness that most of us know as life.

The Buddha spoke the truth. And yet…

Cosmologies

Before we proceed further, let's return to our definition of story as the representation of a sequence of causal events. Because of the word *causal* in that formula, every story opens the door to understanding something about life.

My father died yesterday. We sat shiva for him all night.

That is a story. A man died. This caused his family to grieve in a traditional Jewish custom.

My father died yesterday. The Knicks pulled it out in overtime.

On its face, that is not a story because there is no causation; it is two unrelated events. Ironically, our listener can turn it into a story. We are all so exquisitely primed to hear stories that our imaginations look for the causative links even when there might be none: did the father's death lift a hex on the Knicks? Was this dead man so unloved that his child cared more about a basketball game than his passing?

Because stories carry a kernel of causation, we use them to understand life.

And since we use stories to understand life, perhaps the biggest story of all, the one that human curiosity has always struggled to frame, is how we got here… and what *here* is. The Cambridge Dictionary defines cosmology this way:

The study of the nature and origin of the universe.

Buddhism is the only one of the major religions that doesn't have an origin cosmology, although it does have some ideas about the nature of our world. Science, on the other hand, has plenty of ideas about how our universe came about. A big bang burst about 14 billion years ago. A collection of space trash accreted into an earth about 4.5 billion years ago. Life spawned on said planet around 3.5 billion years ago. Then the process that Charles Darwin first described took over: single-cells to multiple-cells to apes to us.

Almost immediately after I stumbled into the Zendo, I started feeling the urge to meld my practice into the scientific cosmology I held in my head. As to origins, this wasn't difficult because, as I've mentioned, Buddhism is silent about the birth of the world. The challenges came when I faced this question: how can we reconcile survival of the fittest with the Buddha's insight? This was a question outside the bounds of usual Zen Buddhist practice, but I found it impossible to turn away from it.

If you have tried to sit on a cushion – or even if you have not – you know the mind is restless. The Buddhists call this *monkey mind*. Why does the brain chatter endlessly? Why is it so hard to quiet? Why does this organ burn 20% of our body's energy, even when it has nothing specific to do? Surely fitness calls for efficiency; this hardly seems efficient. Is this mental noise adaptive?

Why did evolution mold creatures to be steeped in the dissatisfaction the Buddha talks about? Is *dukkha* adaptive?

If the Buddha is right and life is illusion – if the self itself is an illusion – what evolutionary advantage could possibly arise from living in illusion?

This book is about stories, but it is also an attempt to reconcile the four noble truths with modern scientific thought: to place Buddhist practice within a contemporary Darwinian and materialist* cosmology.

---

* By materialist, I mean that the mind is entirely produced and explicable by physical matter and its processes, with no need for an intervening spirit, soul, or non-physical substance.

Malunkya's Questions

When I write that last sentence, a certain admonition rings in my ears. It is a story from the Pali canon* about a monk, Malunkya, who comes to the Gautama Buddha with some questions of the cosmological variety: Is the Cosmos eternal? Will there be life after death? Not exactly the sort of idle queries I am proffering, but in the same neighborhood.[2]

These questions are wound so tightly inside Malunkya that he issues an ultimatum. If the Buddha doesn't answer them to his satisfaction he will abandon the monastery and his discipleship. The Buddha gets about as ticked off as an Awakened One is liable to get.

"Malunkya, did I ever promise to answer these kinds of questions when you joined the monastery?"

Malunkya admits that the Buddha did not.

"Then, foolish man, why do you act so aggrieved? If a poisoned arrow pierced your skin, and your friends took you to a doctor – would that doctor demand to know who shot the arrow? What sort of bow it came from? What materials the arrow is made from?

"No! The doctor would focus on extracting the poison. Because that is what is important for your life. He doesn't care about the arrow!

"Likewise, I am focused on helping you sever your attachments so you may gain liberation. That is what matters and what you should focus on. What I choose not to speak about, I choose to not speak about!"

Malunkya admits his error and recommits himself to The Way.

---

* This set of texts is the earliest written record of the Buddha's life and teachings.

The Buddha established a methodology of spiritual practice that since his death has allowed countless souls to flower. I've seen it happen.

But – for better or for worse – I am Malunkya. I am obsessed with the arrow.

Let me extend the Buddha's metaphor. An arrow is something linear, with a beginning, a middle and an end. It travels through space effortlessly, in short time. It can carry things: poison, fire, even messages. Story is like an arrow.

But Story is also a tool that became its inventor. If I ask "who are you?" chances are you will tell me a story: where you were born, your family history, the tribes you identify with, when you found (or didn't find) your true calling, what you hope to achieve in the future. Since you are the protagonist of your story, you will tell me about your traits: your strengths and weaknesses.

If I ask you to point to where your strengths and weaknesses are you would have nowhere to point. Those things are just mind matter; yes, they might (or might not) have a correlation with life as you have lived it or will live it, but they don't have a present physical form. We don't just use Story; it is not just a tool. What we call our Self *is* a story.

The Arrow has pierced our soul and our soul has taken the Arrow's form.

The Buddha tells us to not worry about the Arrow – to concentrate on removing the poison instead. I have no doubt he would scoff at the sentence I wrote above:

This book is about stories, but it is also an attempt to reconcile the four noble truths with modern scientific thought: to place Buddhist practice within a contemporary Darwinian and materialist cosmology.

Ha! What useless nonsense.

If your immediate goal is to become more compassionate, to live more in the present, to quiet the torments of your mind, perhaps the best thing to do is take the Buddha's advice. Suck out the poison, don't worry about the arrow. Put this book down and get yourself to a teacher that can start you on a path of practice.

The Buddha spoke the truth.

And yet – and I have no apologies to offer here – in this book I will play the role of Malunkya. I will pose the questions my curiosity demands answers to. And moreover… I will have the gall (which even the chided Malunkya did not have) to say there is value in those questions and answers. There is value to understanding the Arrow.

Diagnoses Can Heal

Studies of the effectiveness of antidepressants reveal a curious phenomenon. As in most pharmaceutical experiments, they are conducted with two cohorts: one that takes the drug, and one that takes a placebo. The truth about who is taking what is hidden both from the subjects who are taking the pills, and the health professionals who are administering the pills.

Studies show that many anti-depressants are modestly, though not dramatically, more effective than placebos, but the interesting thing is that the placebo effect is much stronger for anti-depressants than it is for most other drugs. A large meta-analysis discovered placebo response rates of 35–40%.[3] Why so high?

Let's imagine a depressed person entering a psychiatrist's office. The psychiatrist has got this patter down. She has said it dozens of times before, as she scrawls illegible jumbles on a prescription pad:

Your feelings of hopelessness might be triggered by life events, but you probably have a genetic pre-disposition towards depression. At the most basic level, your emotions are caused by a chemical imbalance in your brain.

I'm going to prescribe a pill that inhibits the uptake of serotonin in your synapses. This medicine has helped a lot of my patients feel better. I think it might work for you.

You can imagine how this might change the patient's outlook. It's worked for others! There's a bright future ahead.

Let's not forget, however, that what the subject of a study on depression hears is slightly different than what the usual patient hears. It's this:

Your feelings of hopelessness might be triggered by life events, but you probably have a genetic pre-disposition towards depression. At the most basic level, your emotions are caused by a chemical imbalance in your brain.

There's an opportunity for you to enter a clinical trial of a new anti-depressant. We don't know for sure whether it works, but this trial will help us find out. There's a chance that you won't receive the drug but instead will receive a placebo. We will monitor your condition and, after the trial, provide continuing treatment. Would you be interested in participating?

The clinical subject enters with doubt as to whether they will receive the actual medicine, and doubt as to its effectiveness. And still, 35-40% of the placebo takers will improve! Why? One explanation might be that depression is naturally more prone to improvement by itself than other conditions. That is plausible and even likely. But another explanation is that it's not the placebo itself that causes improvement – it's the diagnosis.

The patient enters that room thinking they are a damaged and deficient soul. In the psychiatrist's office, they encounter a new narrative. Due to a genetic accident they had no control over, they have a chemical imbalance that makes them sad. This imbalance might be addressable by modern pharmaceuticals. What they have isn't a failure of personality or spirit, it's a small chemical maladjustment that can be fixed – like taking a car to the mechanic.

This narrative is a lot less corrosive to their Self than the one they had before.

This might sound bold, but I hope that this book might – like the Psychiatrist's patter as she is scribbling on her prescription pad – change just a little bit the story you tell yourself about who *You* are. Maybe, as with the people who enter the psychiatrist's office, the diagnosis might lead you to consider a less corrosive model of your Self. And just maybe, understanding your malady might lead you to a medicine that has been clinically proven to have at least as good results as anti-depressants: a practice of mindfulness.[4]

I am not a scientist. I am not an educator or journalist. I am a practicing Buddhist, but not one of much merit or distinction. I am fascinated by ideas, but I can't compare as a thinker to the intellects I will reference in the pages that follow. What I am, what I've always been, is a storyteller.

The story I would like to tell in this book is... the story of Story. It is the natural history of how over millions of generations a being emerged from the primordial pond and developed a number of faculties, each with its own independent fitness value: Consciousness, Self, Emotions, Episodic memory, Mental modeling, Theory of mind, and Language. I will dedicate a chapter to each of these abilities, describing how and why we might have developed them. Finally, I will describe how they combined to create a majestic new skill – storytelling – that gave us the power to shape the world with sequences of symbols.

That is the *origins* part of our scientific cosmology, and it forms the spine of the book. But on each side of the vertebrae that make up the spine, other stories will emerge like ribs; these will describe the *nature* part of our cosmology. One side will be tales of the scientists who, over the last two centuries, forged our understanding of how the mind works. On the other side will be another protrusion of ribs: tales of the monks and mystics of the east who developed a view of the human predicament through a rigorous practice of contemplation. The ribs from both sides will join in a sternum of agreement on this point: that the human mind is multitudinous, interdependent and ever-changing.

Not all the ribs will join! There will be space below the diaphragm for the wind of awe and hypothesis to enter.

Within this torso there might be a heart, but it will be your task to put it there. In this book I will address you directly – I will ask you to imagine, to question, to contemplate. This workbook-like engagement is a convention of some self-help books. I do hope this volume helps you in some way, but there is another reason I am taking this tack: I believe there is probative value in it. In science, introspection is usually not considered a good source of data, but there is only one consciousness that you have access to: your own. No one else is likely to gain access to it, and you are unlikely to ever gain access to others. Science can discover general laws about our species, but with some justification, you probably care most about the law of you. So if I tell you something is true, poke around and see whether your mind works like I say it does. Perhaps, as it was with the sages of the east, contemplation has some truth-revealing power.

So... if you share my curiosity about how we came to exist in this nest of fiction we call Mind, let me tell you a story about how it might have happened.

CHAPTER 1

SOMETHING FROM NOTHING

The superhero protagonist of our tale is the storyteller, and all superheroes need an origin story. How did the storyteller come to be?

To tell a story is to explore causative links, and tracing causative links back to their origin is a very natural thing for a human to do.

"Where do babies come from?" the young child asks.

"A stork brings them," answers the parent, hoping to avoid a premature sex-ed lesson.

"Where do storks come from?"

At this point, the wise course is to suggest to the child that they ask the other parent, but that solution won't do for us. And it wasn't a viable option for a young chemist at the University of Chicago named Stanley Miller, who in 1952 was designing an experiment to explore the origins of life.

Miller asked himself what would happen if you took some of the basic substances thought to exist before the advent of life on earth – molecules like water, methane, ammonia, and hydrogen – heated them, and then exposed these vapors to an electric arc, simulating lightning? What if you repeated this process on a continuous loop?

This might sound like the science project of a high schooler who has seen *Frankenstein* on television too many times, but the results were shockingly fruitful. Within a week, the mixture turned red and cloudy. Analysis of the liquid revealed the presence of 11 different amino acids, the essential building blocks of proteins and life itself. (Decades later, reanalysis of Miller's original samples found evidence of more than 20 amino acids, suggesting his findings were even more significant than initially thought.)

Today, the Miller-Urey experiment (named also for Miller's advisor, Harold Urey) is considered a cornerstone in the study of abiogenesis[5]: the idea that inorganic matter combined to create organic matter, which in turn led to a simple form of life. While scientists still don't know exactly how life began, they've developed a rich array of hypotheses.

One theory posits that life might have originated in deep-sea hydrothermal vents, where hot water released by volcanic activity would provide the combination of movement, heat, and minerals that might have made life possible. Another suggests that ultraviolet radiation from the sun played a crucial role, perhaps in a setting closer to the "warm little pond" Charles Darwin envisioned in his writings. Some researchers theorize that water and fatty acids could have combined to form vesicles—tiny bubbles capable of encapsulating the precursors to life.

The most science-fiction-like hypothesis is built on the discovery of the Late Heavy Bombardment, a period about four billion years ago (just predating the origins of life) where the earth was hit by an unusual volume of comets, asteroids and meteorites. The idea is that some of these celestial projectiles might have carried the seeds of life. One might object that this explains one miracle by proposing two miracles: the beginning of life *and* its transportation through millions of miles of freezing vacuum. But, before we scoff too loudly, we should note that Nobel Laureate Francis Crick, who discovered the structure of DNA, was one of the advocates of this theory.[6]

The question of the origins of life was approached from a different direction by William F. Martin and his colleagues at Dusseldorf's Heinrich Heine University. They aimed to genetically trace the evolutionary tree back to what they called LUCA – the Last Universal Common Ancestor – a hypothetical microbe that might be the grand-daddy of all life as we know it. To do so, they analyzed a database of more than six million genes found in bacteria and other single-celled organisms, searching for shared features.

In 2016, they published their results.[7] They identified 355 shared genes, which collectively painted a remarkably detailed picture of this ancestral organism. Their findings strongly supported the hydrothermal vent theory as LUCA's likely habitat. While other scientists applauded the research, some challenged the conclusion that LUCA was necessarily the origin of all life.

However, even if we gloss over the question of how a bacterium was spontaneously spawned from a seabed or a swamp, another formidable question remains.

Out Of Many, One

For a long time, scientists puzzled over how eukaryotes – a group that includes everything from simple amoebas to plants, animals, and humans – could have evolved from prokaryotes, a category that includes bacteria. The gap in complexity was significant: even the simplest eukaryotes have cells with multiple functional parts, such as a nucleus, mitochondria (which provide energy), and other organelles – none of which are present in prokaryotes. Additionally, eukaryotes reproduce through a method of cell division called mitosis, distinct from the simpler processes in prokaryotes. And most exasperating of all, nobody could find fossil traces of any intermediate form of life.

The person who did the most to solve this conundrum was a fascinating character named Lynn Margulis.[8] From an early age, Margulis was precociously brilliant and precociously willful. She once changed high schools without telling her parents (a bit of forgery was required), to access a richer trove of potential boyfriends, in her own telling. She gained early entrance to the University of Chicago and earned her first degree by the age of 19.

It was at the University of Chicago where she met a graduate student of physics by the name of Carl Sagan, who would gain fame as a science educator on public television. They married shortly after she graduated. This marriage between one of astronomy's greatest communicators and one of biology's greatest theorists only lasted seven years, but it gave them two sons, Dorion and Jeremy.

By the mid-1960s Margulis was an Assistant Professor at Boston University, and she believed she had a plausible explanation for the jump in complexity between prokaryotes and eukaryotes. Maybe the different working parts in a eukaryote – the organelles – were once different species of prokaryotes existing in a mutually beneficial or 'symbiotic' relationship with each other, and eventually they fused together to form a single organism. Instead of the standard Darwinian model of random mutations yielding advantages of fitness that persevered and expanded through generations, she proposed a radical new idea: that communities of organisms had combined to become one in a process called endosymbiosis.

The paper she wrote proposing this hypothesis was called "On the Origin of Mitosing Cells." While it builds on the work of theorists before her, Margulis gave the story a level of detail it did not have before. She marshalled together previous observations that buttressed her claim. And she explained how her theories might be experimentally verified.

Her paper was rejected by more than 15 journals before being accepted by *The Journal of Theoretical Biology*, in 1967. The concept intrigued many scientists and lay people, but it was considered by others to be scientific heresy. A grant organization once responded to her application with the message: "Your research is crap. Don't ever bother to apply again."[9] Remarkably, Margulis stood by her theory even when it was met with ridicule and disdain.

As the years passed, however, evidence for her theory accumulated. In particular, it became possible to sequence the genomes of mitochondria, the energy engines inside eukaryotes, and to confirm that they were distinct from the genomes of the cell nucleus. This suggested that mitochondria had originated as separate organisms, exactly as Margulis proposed.

Margulis was sometimes combative when advocating for her point of view. She once called neo-Darwinians "a minor 20th-century religious sect."[10] Neo-Darwinian Richard Dawkins, who tangled with Margulis at a famous Oxford debate, once alluded to her pugnacity by calling her "Attila the Hen." (Dawkins isn't a timid advocate either.) Still, eventually he had to admit: "I greatly admire Lynn Margulis's sheer courage and stamina in sticking by the endosymbiosis theory, and carrying it through from being an unorthodoxy to an orthodoxy."[11] And orthodoxy it was. While not every detail of her hypothesis has been verified, endosymbiosis theory has such explanatory power that it is included in textbooks even at the high-school level.

Gaia

Before continuing with our story of the eukaryote that becomes a storyteller, let's visit the other idea that Margulis's name has been associated with: the Gaia Hypothesis. Going down this path will lead us back to important questions as to who and what the storyteller is.

Developed in collaboration with British scientist James Lovelock, the Gaia Hypothesis proposes that the sum of life on Earth optimizes the planet's environment for its own use.

This idea annoyed many biologists. *What* is it that was doing the optimizing? Was it some mystical force, they snickered. Lovelock and Margulis insisted that they were *not* proposing that the biosphere was a single living organism, although their ideas were sometimes interpreted that way.

Lovelock believed that no mystical force or singular organism was needed to produce the dynamic equilibrium proposed by Gaia.* A glance over to the field of economics shows how this can happen. Firms produce more of a product as its price increases (supply), while consumers buy less of a product as its price rises (demand). The equilibrium price is found where these two forces intersect. Similarly, nature abounds with such interdependent dynamics. For instance, predator populations tend to decrease as prey populations dwindle, and prey populations tend to rise as predator numbers

---

* Unlike Margulis, Lovelock wavered in how he talked about Gaia. When speaking or writing for scientific audiences, he would adopt a more disciplined posture, but when speaking to a lay audience, he would sometimes refer to the earth as "a kind of organism."

fall. Where the two lines intersect is where the population balance will gravitate toward. Lovelock and Margulis observed that there are many such equilibriums in our natural world, and that these relationships don't just involve organic matter but also the inert solids and gases that life takes in or excretes. There is an Invisible Hand, as the economists say, but no mystical will. Lovelock later illustrated this concept with a computer model called Daisyworld, which demonstrated how simple feedback mechanisms could maintain planetary stability.

Seen in this light, the Gaia Hypothesis seems less like a bold conjecture and more like a simple, apparent observation.

Margulis opens her book *Symbiotic Planet* by recounting a question her son Zach once asked her: "Mom, what does the Gaia idea have to do with your symbiotic theory?"[12]

Margulis answers: "nothing, or at least nothing as far as I'm aware." But she also approvingly quotes a student of hers, Greg Hinkle, as saying: "Gaia is just symbiosis as seen from outer space." Perhaps Margulis's denial of any connection between the two ideas carried a hint of disingenuousness.

For me, the similarities between the two ideas are clear. Both symbiosis and Gaia relativize the categories of what an organism is. I tread carefully here because Lynn Margulis would probably flinch at my impure motives. She wrote about the reception Gaia got in some quarters:

Environmentalists and religiously inclined people, attracted to the idea of a native goddess with power, latched onto it, giving Gaia a distinctly non-scientific connotation.[13]

I will protest: I suppose I am religiously inclined, but I attach no goddess identity to the biosphere. I'm empathetic to any scientist who might find my company embarrassing, but I think my position is worth explaining further.

Before we can answer the question: "Is the earth an organism?" we need to ask: "what is an organism?" And before we ask that, it might be wise to consider: "what is a *category* such as organism?"

Categories And Beyond

Take any set of items, and you can categorize them in different ways depending on your purpose. For example, consider the set of road vehicles.

A car dealer might divide vehicles into Used and New. The new cars must be ordered from the manufacturer; the used ones have to be purchased privately and reconditioned.

A highway engineer might divide cars by weight. You need a tougher road for heavier vehicles.

A policeman might look for the most immediately identifiable markers: color and make. You can't catch the bad guys if you can't make a quick ID.

These categories are not arbitrary, they are not useless – on the contrary, they are invaluable cognitive tools. But they are mental and social constructs: different ways to slice up the same set.

Lynn Margulis once wrote: "I cannot stress strongly enough that Gaia is not a single organism." Well, before we can say whether we agree or disagree, we need to ask: what is the definition of an organism? Surprisingly, there's no universally agreed-upon formulation, but organisms typically share certain characteristics. The earth does possess some of these characteristics: it is adaptive, it maintains homeostasis (a dynamic equilibrium) as Lovelock and Margulis observed, it contains genetic material – among other things.

However, there are other defining characteristics that the earth doesn't have: the ability to reproduce, for example; the world has so far not created other worlds. Likewise, organisms excrete waste outside their periphery; here on earth we keep our wastes right where we are. These differences help explain why Margulis resisted categorizing Earth as an organism.

But let's imagine a hypothetical: say Elon Musk whisked us off to Mars to start a new civilization, and now we can say the Earth reproduced. Say we started to dump radioactive waste into cylinders that were rocketed into outer space, and now we can say the Earth "excretes." Say we covered all the possible objections ... would moving to a new category change the earth's essence? Or to tweak Shakespeare, would a rose by any other name smell any different?

Is the following statement true or false:

You and I are organisms, but we are also cells in a larger organism: the Earth.

The answer depends on how you define the category of organism. As we asked with the vehicles, we must ask: what is the function of the categorization? To say we are trying to better our understanding is not helpful – it is too vague. An environmentalist trying to persuade us to forge a more respectful relation with the globe might be a bit more loose about the definition of 'organism' than a microbiologist like Margulis, who has her own everyday use for the term. Can we say that the environmentalist is wrong about a construct that has been defined in so many different ways? Ultimately, the validity of the statement depends less on inherent truth and more on the perspective and purpose behind the categorization.

A category is a construct, a tool to help us think, but it also shapes the *way* we think. Margulis understood this well. Another great project of her life was taxonomy, the categorization of life. Taxonomy doesn't just name organisms; it also tells the story of how species evolved from one another. The metaphor commonly used for this process is a tree, with species branching off over time. Since Margulis was putting forth a bold notion with symbiosis, that these branches of species not only divided but merged sometimes, she preferred a different metaphor: that of a web. Taxonomies help us name, but they also help us understand, so Margulis preferred a system that was apt for her view of the world.

Unfortunately, taxonomies – and categories – also help us misunderstand. We don't want to be highway engineers working with the categories of car dealers, or policemen working with the categories of highway engineers.

When it comes to evolution, there are multiple mechanisms at play. Margulis herself sometimes became too attached to the category of symbiosis and too skeptical of the role of natural selection. She wrote, "I believe that most evolutionary novelty arose, and still arises, directly from symbiosis." But does the evidence fully support this claim? While bacteria can exchange genetic material horizontally, the kind of organism-fusing event that created eukaryotes seems to have been a freak occurrence – a once-in-many-eons phenomenon. Was Margulis favoring an explanation that gave her a central role as a pioneering theorist?

To her credit, she did not lack self-awareness:

Some colleagues label me combative; others, unfair. Some say I only collect relevant work and unfairly ignore contradictory data. These accusations may be correct.[14]

In 2009, one year after Françoise Barré-Sinoussi and Luc Montagnier were awarded the Nobel Prize for identifying HIV as the cause of AIDS, Lynn Margulis proposed an alternative theory: that spiral-shaped bacteria called spirochetes, living as symbionts in humans, were responsible for the disease[15]. This suggestion came despite the lack of empirical evidence to support the presence of such organisms. Was her enthusiastic commitment to symbiosis as a framework for understanding biology pushing her toward something close to a conspiracy theory?

Not Knowing

Our foray into the question of "what is a category?" leads us right back into the heart of Zen Buddhism – quite literally, the seminal bit of scripture we call the Heart Sutra. In Zen, we try to let go of categories in our thinking. We call this not-knowing. To get to the place where we can detach from our cravings, we need to *not know.*

Not know that this is beautiful and that is ugly.

Not know that this task is pleasure-giving, and that task is annoyance-creating.

Not know that this person is charming and intelligent, and that person is awkward and dull.

When we quiet our minds in meditation, we quiet the stories that are spinning in them.

When we quiet these stories, we free our minds from knowing categories like beauty and ugliness, pleasure and annoyance, charm and dullness.

And when we free our minds from knowing, we free

ourselves to experience the world freshly and without preconception – we break free from the inner taunts and lashes that knowing submits us to.

In this world where knowledge is prized as a currency that defines status, that can gain you advantage in both professional and social ambits, it might seem strange to posit not having knowledge as a preferred state. But to not know categories is an important part of what the Heart Sutra calls emptiness.

Oh Shariputra, form is no other than emptiness,
Emptiness no other than form;
Form is exactly emptiness, emptiness exactly form.

The sutra goes on to assign emptiness even to the categories that define Buddhist practice:

No realm of sight, no realm of consciousness; no ignorance and
no end to ignorance,
No old age and death, no end to old age and death,
No suffering, no cause of suffering, no extinguishing,
no path, no wisdom and no gain.

Buddhism insists that there is no Buddhism – no categories at all.

Freeing your mind from categories can free your mind to expand. Margulis's great contribution to science was taking a giant leap outside of the established categories of thinking. Later in her life, categories sometimes seemed to shrink her great mind.

Of course, it is impossible to live life with no reference to categories. You might leave home with shoes as gloves and underpants as a hat. But to make a practice of resetting your mind to zero categories is not the least bit wooly-headed. On the contrary: it is an attempt to arrive at a baseline for truth.

This idea is important for scientists because the scientific method involves coming up with a hypothesis and testing it. Laurels are rarely given to those whose hypotheses fail to be validated. Margulis herself was fond of quoting the physicist and philosopher David Bohm as saying: "Science is the search for truth... whether we like it or not."[16] That's admirable, but perhaps by the time you see whether you like the results it is already too late. Perhaps by then a supposition underlying the experimental design has slanted a trial in a certain direction, or perhaps an uncooperative data point has been excluded for a plausible but dubious reason.

You cannot search for objective truth without interrogating the subjective.

The Edge Of You

How about you? Where do you start? And where do you end? Is it at the edge of your skin, or is it at the edge of the atmosphere?

Are you part of a larger whole? While the idea is hotly debated, many evolutionary scientists support the concept of group selection – the notion that behaviors benefiting the group can provide a fitness advantage for the group, advancing those traits forward, even if they involve some degree of sacrifice. Is your tribe an organism? Your workplace? Your nation?

How about the bacteria in your gut? You and they are, in Lynn Margulis's terms, true symbionts. You provide them with a warm, nutrient-rich home; they help you break down complex carbohydrates and absorb nutrients. A healthy life would be almost impossible without them.[17] Are they part of you?

If you say they are not you because they could live life outside of you, isn't that true for many other transplantable cells in your body?

If you say the bacteria aren't you because they act in their own self-interest, how do you define that? After all, not a single cell in your body has the will to contribute to the project of you.

If you say the test of you is whether the cells carry your specific DNA, then that means that most of your blood cells (which don't have a nucleus or DNA) would not be you, either. One of the discoveries validating Margulis's theory was that the power plants inside your cells – the mitochondria – have a different genome than the nucleus cells. Does that mean that a part of you isn't really you? Or taking it further, if your identical twin has your identical DNA, does that put him or her within the edges of you?

Are your nails you? Nails are made of dead cells. Can they be dead and also you? If you clip your nails, are they still you as they flush down the toilet? How about if you paint on them? Is the paint you? Nail paint is a very intimate form of self-expression. If you paint them red before you go on a date, it might mean one thing. If you paint them black before going to an emo concert, it might mean something else. Maybe the paint is more you than your nails are.

If categorization follows function – as with our car categories example – what is the function for the category of you? Whose interests do you want this category of you to serve? Taxonomists? Lawyers? Buddhists? You? (Never mind how circular that might be.)

Now notice your breath for a few seconds.

Where did the categories go? Where do they exist?

Just in your head, you say? If you're not thinking of them, do they still exist?

And what happens if you stop thinking of the category of you?

CHAPTER 2

PARLIAMENT OF MIND

Lynn Margulis helped us understand one of evolution's greatest leaps: how independent organisms merged to form the first eukaryotic cells, the building blocks of all complex life on earth. This was a biological revolution, but it was only the beginning. At some point, life forms became sentient – capable of experience, memory, and cognition. Stories are the sharing of consciousness – this noisy, miraculous, multi-media rave that is going on in our heads right now. How did we get from amoebas to ideas?

Tracing this journey requires many guides, but the first and most obvious is Charles Darwin. When *On the Origin of the Species* was first published in 1859, Darwin was already something of a minor celebrity. His earlier book, *The Voyage of the Beagle,* had been well received by both scientific and general audiences. As a naturalist, he had made important contributions to the understanding of coral reefs and barnacles, among other things. And he had a wide network of scientific colleagues that he corresponded with, exchanging evidence, opinions, and impressions. The proposition of his theory of evolution in *On the Origin of the Species* had an immediate impact on both specialist and lay audiences.[18]

During that same year, a continent away in what is now the Czech Republic, a German-Czech monk by the name of Gregor Mendel was busy conducting a series of experiments on peas. Mendel identified seven characteristics in these plants – such as color and form – and crossbred them to observe how these traits appeared and disappeared over successive generations. What Mendel uncovered were the basic laws of heredity, but he never achieved the acclaim during his lifetime that Darwin did. Mendel's accomplishment was seen as a finding of limited applicability to the field of agronomy, and nothing more. The true significance of his discoveries wasn't recognized until the turn of the 20th century, more than a decade after his death, when three separate research teams independently rediscovered his work.

This delay was unfortunate because Mendel's findings filled a crucial gap in Darwin's theory. Darwin wrongly hypothesized that every part of the body created what he called gemmules, and that these gathered in the reproductive organs and combined with their pair's gemmules during reproduction. Mendel, on the other hand, introduced the idea of "factors" – a term later replaced by "genes" – and described how these interacted. It was Mendel who revealed the mechanics of dominant and recessive traits.

During the first six decades of the 20th century, Darwin's theory was merged with Mendel's discoveries and enriched by the discovery of DNA's double-helix structure. This integration became known as the great Neo-Darwinian synthesis. While this framework has been challenged and refined by theorists such as Margulis, the basic architecture still stands.

This architecture has never been more elegantly described

than in Richard Dawkins' 1976 classic *The Selfish Gene.** Dawkins chooses to start not with the concept of factors or genes or traits, but with the concept of a Replicator. Evolution doesn't even require life, he writes. It just requires a Replicator that can reproduce itself with some genetic novelty introduced occasionally by way of mutations.

Our ancestral Replicators had one thing in common: there were things that were good for them, and there were things that were bad for them.

Nourishment was good. Toxic elements were bad.

Potential mates were good. Predators were bad.

The Replicators who survived were those who mutated in ways that allowed them to steer toward the good and away from the bad; they passed this ability on to their progeny. The Replicators who couldn't do this perished, and their less fit bloodline perished with them.

Into Consciousness

How did we get from single-cell replicators to that thing we call consciousness? Before I describe how this might have happened, let's clarify what we mean by the term consciousness, since it is used in so many ways. I do not mean self-awareness or any other higher capability, but rather the ongoing combination of sensory information, cognitive processes, and emotional activity that is happening within you right now.

At first, Replicators had no need for consciousness. They used simple, modular mechanisms that operated in a reflex-like, discrete manner.

Sense the scent of food? Move towards it!

---

* Dawkins' emphasis on competition at the genetic level, however, has been criticized by some scientists as underplaying other evolutionary factors.

Sense the scent of predators? Move away from it!

But over time, organisms needed mechanisms to mediate between conflicting impulses. What if there was both the scent of food and the scent of a predator? What should the organism do? At first these mediating mechanisms were likely straightforward and rule-based:

If there's a predator, forget the food.

As these modular impulses become more numerous and nuanced, animals needed to develop central nervous systems to coordinate cohesive responses to their environment. Sensorial channels of information were pooled together to create a sophisticated facsimile of the world in their mind, so that signals of food or danger became more complete and more reliable.

To illustrate this, imagine that a monkey in the jungle has caught the aroma of a ripe mango fallen from a tree. It is the promise of a feast! But then, he sees some nearby bushes rustle. It could be a predator – or just the wind blowing the branches. The aroma tempts him forward, but then he hears the unmistakable sound of twigs snapping under the weight of a feline paw. That's no wind; it's a predator. Instead of grabbing the mango, the monkey climbs a tree for safety. The mango can wait; safety comes first.

In this moment, the monkey's consciousness is combining channels of sensory information. The sight of rustling leaves is a weak clue that a predator might be nearby; when combined with the sound of snapping twigs, however, it provides near-certain proof of danger. If the monkey processed each signal in isolation, without connecting and evaluating them collectively, he wouldn't be able to come to a sound conclusion.

But consciousness does more than just merge sensory inputs. It also provides a space for the monkey to weigh competing modular impulses. What is more urgent: the fear of a lurking predator or the hunger for a ripe mango? Has he been in this situation before? Survival depends not only on gathering sensory data but also on having a workspace in which to assess and act on that information.

Importantly, this pooling of sensory and cognitive information didn't eliminate the modularity of mental responses. Instead, modular inputs were integrated into a unified consciousness, enabling the organism to adjudicate between competing or ambiguous signals.

This evolutionary story aligns with Leda Cosmides and John Tooby's Massive Modularity hypothesis. It also dovetails with the Global Workspace Theory, which conceptualizes consciousness as a central hub of information exchange. Such a framework helps explain one enduring mystery about consciousness: its near-indestructibility. Of course, we can fall into coma states, but human consciousness is shockingly robust. Brain injuries and lesions rarely extinguish it entirely, and there doesn't seem to be a single location in the brain responsible for generating it. Consciousness is like a public address system that fills our whole head, which is exactly what we would expect if its function was to share information.

One possible objection to this theory is to ask why such computation needs the light of consciousness to occur. Couldn't all these operations happen algorithmically – in the dark, so to speak, without our awareness?

Our consciousness *is* the algorithm, it *is* sensorial information pooling, it *is* the neurons firing. To argue that it cannot be so – simply because we *feel* our consciousness – is to fall into the illusion of subjectivity. Yes, there is a part of you with the ability to survey and consider your thoughts, but that doesn't mean that there is a little person inside you whose presence gives you the magical power of consciousness. That "little person," that module or neural network, is just an onlooker – not a conjurer. We shall say more about the fallacy of the little person below and in further chapters.

But coming back to modularity – what evidence is there that our brains are comprised of specialized modules with distinct functions? A case study more than a century ago laid a cornerstone in this theory.

Time Time

One day in 1839, a 30-year-old Frenchman named Louis Victor Leborgne walked into Bicêtre Hospital in the southern suburbs of Paris.[19] Leborgne had struggled with epilepsy since his youth, but the latest complication was an alarming one. Leborgne had completely lost the ability to speak.

In its stead, Leborgne voiced a single syllable that he uttered with different pitches and intonations, but always with the same sound: "Tan." Usually, he would use the syllable in pairings: "Tan, tan." In French, this syllable sounds most like the word *temps*, or time, so this unfortunate man – condemned to inarticulation – might still have seemed to the staff to have some poetry in him. They started using his preferred sound as his nickname.

The remarkable fact of his ailment was that the rest of his intellectual functioning seemed largely unaffected. Leborgne was described by those who dealt with him as intelligent.

Sadly, his father – a schoolteacher – died shortly after his admission to the hospital. His mother had died when he was three, and he was unmarried. Severely handicapped and with no caretakers available, he would remain in Bicêtre for the remaining 21 years of his life.

Toward the end of his life, Leborgne's condition deteriorated. He became bed-bound because of a paralysis. Bed sores on his right leg developed into gangrene.

Attending him was a brilliant young surgeon named Paul Broca.

More than just a surgeon, Broca's broad-ranging curiosity led him to make important discoveries regarding rickets, arthritis, and muscular degeneration, as well as to venture into far-off fields such as anthropology.* However, Broca's skills were not enough to save Leborgne's life. In 1861, Leborgne succumbed to the blood poisoning that gangrene causes.

It would be uncharitable (and unfounded) to say that Broca was hoping for Leborgne's death, but it's hard to imagine that there wasn't some anticipation. Not long before, Broca had seen a presentation by Ernest Aubertin at the Anthropological Society of Paris. Aubertin's story was both gruesome and fascinating. A young man had come to Aubertin's hospital with an open gunshot wound caused by a failed suicide attempt. The exposed brain contains no nerves for feeling pain, and the history of neurosurgery is filled with tales of doctors poking at the brains of their waking subjects and observing the results. This was one of those cases.

Aubertin applied light pressure to an area of the left frontal lobe that some theorists had suggested might control speech. His action cut off blood supply to that part of the brain. The patient, who had been talking freely suddenly fell silent. Upon

---

* Alas, his curiosity also led him to work up many now disproven theories, such as the idea that there is a relationship between brain shape and intelligence.

releasing the pressure, his ability to speak returned; it was as if an on/off switch for speech had been flipped. Aubertin saw this as compelling evidence that this part of the brain was dedicated to speech.

Impressed by this presentation, Broca had asked himself: *Could Leborgne's malady be caused by a lesion in that same area?* He now had an opportunity to test that hypothesis. Broca performed Leborgne's autopsy and was able to confirm that his hunch was correct. There was indeed a lesion in the third frontal convolution of the left hemisphere, which we now recognize as the center of speech.* Broca did not tarry in reporting his results: he made a presentation to the Anthropological Society the very day after Leborgne's death.

That area of the brain is now known as Broca's Area, and Leborgne's malady is referred to as Broca's Aphasia. Two years later, Aubertin wrote a paper documenting the theory and evidence that had prefigured Broca's discovery,[20] but alas, the glory had already gone to someone else. And so began a long line of inquiry that continues to this day: the effort to localize behavioral functions in specific parts of the brain.

Leborgne has had many cousins-in-science. Phineas Gage's personality changed when an accidental explosion drove a metal rod through his frontal lobe. Henry Molaison lost his ability to form new short-term memories when a surgeon removed about two-thirds of his hippocampus. (We shall return to Molaison and his surgeon – after which I am named – in Chapter 5.) Oliver Sacks wrote about the case of an anonymous male who developed hyper-sexuality after a part of his brain was removed to treat his seizures (the same clinical

---

* The exact location of Broca's area can vary depending on which side of the subject's brain is dominant, and other factors.

motivation that led to Molaison's surgery.)

Modularity

Cases like these, where specific mental functions are altered when the brain is altered, have ravaged the idea that mind is ruled by an ether or spirit, as Descartes thought. And they buttress the case for modularity.

Indeed, the concept of modularity has gained broad acceptance among scientists and philosophers, although it is a house with rickety foundations. Steven Pinker can write a book titled *How the Mind Works* and Jerry Fodor can reply with *The Mind Doesn't Work That Way*; evolutionary psychologists Cosmides and Tooby can hypothesize about massive modularity, and many researchers since can respond with carefully considered rebuttals; but they all usually believe in some sort of modularity, broadly defined. It is difficult to controvert that cognitive functions exhibit some level of independence. The real debates lie in defining how much independence exists, what form it takes, and how hierarchical structures fit into this framework.

Are horrible accidents and unfortunate surgeries the only evidence we have of modularity? Far from it. Since the 1990s Functional Magnetic Resonance Imaging (fMRI) has allowed us to identify what parts of the brain are working by capturing images of blood flow. Using this technology, scientists have observed brain activity during specific tasks and demonstrated that certain tasks activate consistent, specific networks (with some variability) across a wide range of individuals.

This doesn't mean that there are specialized areas in your brain for each task you perform; the mind is far too complicated for that. Instead, fMRI shows that nearly any task is going to spark a network that spans many parts of the brain.

Crucially, these networks are consistent and repeatable: the same ones activate when the same task is performed. Whether you're experiencing lust for a sexual partner, recognizing a loved one's face, or feeling fear in the face of potential danger, specific neural networks –assigned specific functions – are at work in your brain.

The case for modularity does not depend on physical examination alone; it has been supported by behavioral tests as well. The following is called the Müller-Lyer illusion. Which of these lines is longer?

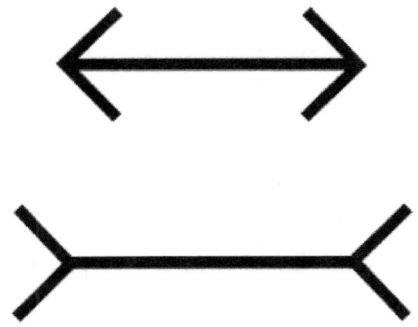

You have probably replied that the bottom line seems longer. If you are enterprising and brought a ruler to the page, you would see that the lines are actually the same length. But does that discovery (from either the ruler, or reading the last sentence, or both) make the bottom line stop *seeming* longer? No, the illusion prevails. It's as if there were a department of visual perception in our brain that has been tricked up, while a department of knowledge in our brain that has gotten an email saying "don't trust what the visual perception department says; both lines are the same length." If there were no wall between the two departments, visual perception might correct its mistake upon learning of it, but instead it continues in error.

If our minds consist of teeming hordes of mental modules, how do we reconcile that view with our experience? Sitting here, laptop on my lap, looking out the window, I can't say that my experience *feels* modular. The input from my senses come as a seamless, uninterrupted flow. My mind has a voice that feels… like the voice of that "little person" we spoke of before.

Is there a boss of my mind that rules over the unruly modules?

A long line of philosophers have pointed out the theoretical shortcomings of what is called the *homunculus* model – that there is a diminutive character standing on top of our minds, steering us. If such a figure ruled our mental processes, the obvious question arises: who rules *them*? Is it like the Hindu myth of the world resting on the back of a turtle, which itself rests on another turtle, and so on? Is it turtles – and homunculi – all the way down? Clearly, this idea collapses under scrutiny.

Furthermore, experimental evidence suggests that our conscious experience may actually *lag behind* our decision-making.

For instance, neuroscientist John Dylan-Haynes and his colleagues at the Max Planck Institute in Leipzig, Germany, conducted a groundbreaking experiment using functional magnetic resonance imaging (fMRI).[21] They asked 14 participants to make an arbitrary decision while in the scanner: press a button with their right hand or with their left hand. The subjects were also instructed to identify the exact moment when they made their choice by reporting which letter of a flashing sequence they were seeing when they made up their minds. The researchers were able to tell *which* button the subjects would pick but, most remarkably, they were able to detect this neural evidence up to 10 seconds before the subject consciously *thought* they were making the decision.

So, it seems some unconscious process makes the decision – and a conscious process later claims credit for it. Is the "boss" in my mind really in charge, or does he just *think* he's the boss?

Findings such as these are not isolated. In fact, there are loads of experimental results that show some form of cogitation or decision-making occurring before consciousness of it appears (although it is usually by a matter of milliseconds rather than the full seconds reported by Dylan-Haynes and his colleagues).

To complicate things further, it's a mistake to think of consciousness as a linear process, like film passing through a projector's gate. There is no moment of truth when the projector's light hits the film and we are conscious of a decision or a sensorial input. Consciousness is more like a constantly revised set of movie frames that *seem to us* like a projected film. (This is what philosophers such as Daniel Dennett have called the multiple drafts model of consciousness.)

Michael Gazzaniga has conducted many experiments on patients with a transected brain:[22] that is, patients whose left and right lobes of the brain were separated with an incision in order to prevent seizures. It is a testament to our brain's resilience that these patients maintain most of their normal functioning. They also present a unique opportunity to conduct experiments with some profound philosophical takeaways.

In one of Gazzaniga's famous experiments, a split-brain patient was shown a card with the word "WALK" presented only to the left eye's visual field, which is processed by the right hemisphere. Since the right hemisphere lacks access to the brain's language-processing areas in the left hemisphere, the patient could not articulate seeing the word. The right side of his brain was able to process this message – perhaps recognizing it in a procedural, non-verbal way since we see a WALK sign every time we cross a street. The patient stood up, presumably in response to this WALK suggestion.

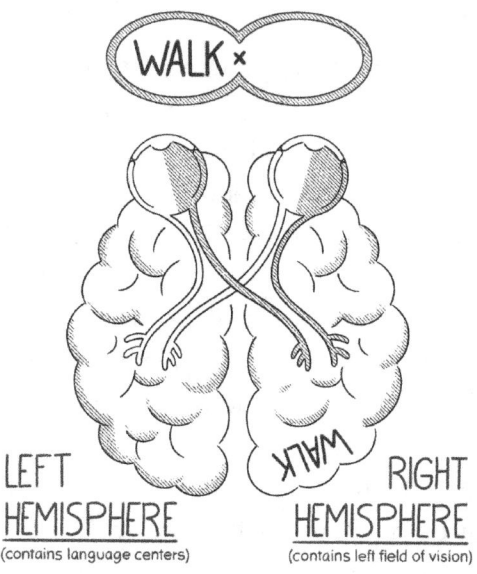

LEFT
HEMISPHERE
(contains language centers)

RIGHT
HEMISPHERE
(contains left field of vision)

When the experimenters asked what he was doing, he replied "I'm going to get a coke." The articulating part of his brain was not aware of the WALK signal, but made an after-the-fact rationalization of the decision to stand and walk for the experimenters' consumption. Once again, we see the conscious "boss" claiming credit for a decision it didn't actually make. Is the boss really the boss?

We don't need elaborate scientific experiments to recognize the existence of unconscious mental processes. Have you ever woken up with a solution to a problem that was unsolved when you went to sleep? That problem-solving did not require your consciousness to occur. History offers striking examples of such sleep-bound creation. As James Kingsland recounts, Dmitri Mendeleev is said to have dreamed up his periodic table of elements in his sleep.[23]

We don't even need to be asleep or unconscious to recognize cogitation happening beyond our view. If we play a musical instrument or we simply speak a language, we can recognize complex lexiconic decisions being made without our being conscious of them.

The Deluded Monarch

As I have before and will again, I invite you to an act of imagination. This will be an explanatory metaphor and, of course, all metaphors break down under inspection at some point. If they perfectly mirrored reality, they wouldn't be metaphors.

So I ask you: imagine your mind as a Parliament of many modules.

There are many unruly back-benchers in your Parliament: louts yelling for sex, food, power. Fearful representatives screeching about possible peril at every instance.

There are specialist technicians: a language specialist is over there scribing by the well of the Parliament; a visual perception specialist is unfurling an exhibit; a memory specialist is filing things away in a card catalog in the corner.

There are multi-taskers that can take on a number of different projects. They are the Parliament's utility players.

There is a Cabinet of the Self that upholds a vision of your organism's place in the social network: a Minister of Biography, a Minister of Aspiration, a Minister of Self-Appraisal, a Minister of Social Affiliation, and others.

There is an Inhibitory Veto Desk where dangerous or destructive proposals go to die.

I will leave the vast argument of how the Parliament actually comes to behavioral decisions to more able and informed minds than mine. Suffice it to say, it probably isn't an actual vote. It's probably a combination of who is yelling the loudest, which coalitions are joining strengths, and what hierarchies have the upper hand. But however it happens, the Parliament of the Self issues a decision. In this case, it's: *yes, let's accept Jeff's invitation to lunch.*

And out onto the well of the Parliament steps a kingly figure with a regal bearing. One of his aides places a piece of paper in the King's pocket without him noticing – it is the Parliament's diktat. His retinue leads him to the Royal Balcony, where the curtains are drawn back so that he may step outside, to make an announcement to the world. With noble haughtiness, he takes out the message from his pocket, which he distinctly remembers having written himself:

"I have decided that I shall accept Jeff Rosenwall's invitation to lunch."

The Monarch, of course, is deluded. The Monarch thinks he is running things, but really the Parliament is. The Deluded Monarch is the process in our mind where a decision enters into our awareness and we ascribe it to *our will*.

That we have a Deluded Monarch is why *homunculi* theories and Cartesian ethers are so easy to imagine, and materialist computational schemes are so hard to imagine, much less credit as being capable of consciousness.

Why have a Deluded Monarch? Consider the counterfactual. If we didn't have a Deluded Monarch, what would our experience be like? We would be exposed to the chaotic din of the Parliament of Mind, sure. It might feel a little like a schizophrenic breakdown.

But more to the point of social fitness, how could we present ourselves to others as reliable social partners if we didn't speak with a single, persisting voice? Just as how a Secretary of State has to ignore the competing and often conflicting interests of her country's constituencies, and speak to foreign diplomats with a unitary view and a unitary policy, so our Deluded Monarch is there to represent us in a consistent way. Survival in our ancestral environment – and even more so in the complex social world we inhabit now – depends on our ability to form pacts and alliances.

Our Deluded Monarch not only helps us present a cohesive, convincing front to our social partners; he or she can also warn of conditional responses or even make implicit or explicit threats to our enemies, rivals, or potential antagonists.

Apart from the outward, social benefits, there may be some self-regulatory benefits to having a Deluded Monarch. By having the marker of a Deluded Monarch's balcony address, we can regulate our impulses so that they serve our organism's interests. Hearing ourselves make the royal announcement that we intend to do or seek something may help us to commit to that action. It may let the modules know that the matter is closed for the moment.

King Milinda's Chariot

The Buddhists tell a parable about a deluded Monarch, but he is not quite as much the butt of the joke as in my tale.[24]

King Milinda, having heard of a wise Buddhist monk, loads up his chariot, gathers his retinue, and sets out to meet the old man. Upon arriving, he asks the monk, "What, sir, is your name?"

"They call me Nagasena," the monk replies, "but that is merely a convenient name. There is no such individual here."

"That's absurd," Milinda says. "If there is no monk, then who have I come to see? If you don't live a righteous life, and a murderer does not live a terrible life, then there is no good or bad, is there?"

Nagasena purses his lips and tilts his head.

"If there is no you, what is this bag of bones, skin, and hair in front of me? Is that not Nagasena?"

"Not really," Nagasena replies.

"I don't believe it," the King mutters.

"How did you come here?" Nagasena asks.

"Why, in a chariot."

"Is the chariot its wheels?"

"No, not just its wheels."

"Is the chariot the rope that leads from your servant's hands to your horses?"

"No, of course not," the King protests.

"Is it the axle? The chassis?"

"No! Again, no."

"Then you have not come in a chariot," pronounces Nagasena.

"Of course I have come in a chariot! The word Chariot is what we use to describe that collection of things," says the King.

"Exactly," agrees Nagasena. "In that same way, I am a collection of my aggregates. There is no unified Nagasena, no permanent Nagasena."

The King is satisfied with this answer. "If the Buddha himself were here he could not come up with a better reply!"

Similarly, a 21st century neuroscientist might be hard-pressed to find a fitter metaphor.

Even though we ascribe our actions to our will, there is no single will, no permanent protagonist of our tale. Our organism's many discrete responses to our environment have been integrated into a cohesive network, but each response maintains its own function and character. We are all aggregates of our mental modules.

This was not an intuition that came exclusively to the east; famously, David Hume came to a similar conclusion in the 18th century when he wrote his *Treatise of Human Nature*.* For Hume, as for the Buddha, the self was an impermanent, ever-changing bundle of perceptions.

Revolutions

One last royal tale for this chapter!

In the 1640s the English had a nasty spat between a Parliament that felt it deserved a little deference and a King, Charles I, who believed he had a God-given right to rule. A civil war ensued and, in 1649, Charles I was beheaded outside the Banqueting House on Whitehall. For a time thereafter England would be ruled by chaos and puritanical forces.

Eleven years later, the son of Charles – Charles II – was invited by the English Army to return to London and reclaim the throne, on the condition that he accept some limitations to his powers. He did so, and thus began a period of Restoration where two decades of violence, intolerance, and instability subsided, and the arts and sciences blossomed again.

---

* In a curious coincidence that I will, with great restraint, not make too much of, Hume wrote this tome while living in the town of La Flèche in France. La Flèche translates to English as "The Arrow."

I won't offend the Stuarts by saying that either Charles was deluded or not deluded. I will say that, just as there was a Charles I and a Charles II, we have the opportunity to have a Monarch I and a Monarch II.

Monarch I stands on the balcony and reads the edicts thinking they are his decisions, and his decisions alone. He is our Deluded Monarch. He believes that all the power to rule his organism resides in his Royal Self.

Monarch II stands on the balcony and reads the edicts, knowing they are the handiwork of many members of Parliament. Monarch II recognizes and honors the diverse quality of his Self.

Such a personal revolution is possible. Why is it important?

To live as if our Self is unitary is to potentiate the idea that we have a persistent and immutable character. Moreover, ascribing a persistent and unitary character to ourselves does three things.

First, we are both too kind and too cruel to the All-Powerful Monarchs who think they rule us. Sometimes we pump up their powers – *aren't they good? Aren't they kind? Aren't they smart?* (Which of course is to think: *I'm good, I'm kind, I'm smart.*) It's a happy ride up the hill.

But on the other side, when they disappoint us, we turn on these Royals with a fury that makes for a rattling ride down the hill. *Why are they so awkward? Why do they say such dumb things? Why are they so cold and mean?* (Which is to say: *why am I so awkward, so dumb, so mean?*)

Monarch I is the peg upon which stories of both narcissism and self-loathing can hang.

Second, not recognizing the multiplicity of the Self makes it easier to suppress parts of yourself.

Say you are a young person looking for a vocation. Should

you be a poet or a lawyer? Should you choose social work or sales? When faced with such choices you will probably have several parts of yourself screaming for attention: the risk-averse you, the please-your-parents you, the fame-and-fortune-seeking you, the idealistic you, the artistic you, the materialistic you, etc. You will encounter people or ideologies that will tell you to suppress some of those yous. They will want you to act as if some parts of you don't exist or are not legitimate. It is a mistake to listen to them. The parts of you that they think you can suppress will persist and will have to co-exist with your decision. Denial is never a promising strategy for navigating life. We are best off if we acknowledge our diverse needs, fears, and desires.

Third, recognizing your multiplicity makes it easier to foster change. Change is no longer a small miracle; instead, it is just another part of you coming to the fore.

So if we commit to this subtle revolution, how might we install a Non-Deluded Monarch II? Unfortunately, having an intellectual understanding of modularity is not enough to erase the deep ruts that run through our mind, our persistent habit of bowing to a Monarch we imagine as ruling us, our habit of seeing ourselves as unitary rather than multitudinous. To achieve that transformation requires commitment to a program of self-observation, which I will outline in Chapter 10.

Installing Monarch II is not the end-objective of our practice. Realizing the diverse quality of our self is just a step in discovering our true nature – what in Zen we call our Original Face. However, our version of the English Restoration is an important step to loosening the tyrannical power that Story has over us.

Next, let's take a closer look at how we developed a concept of our Self.

CHAPTER 3
THE ATOMIST SELF

We've talked about how life might have come about, how consciousness might have come about, and we've urged you to be skeptical of Deluded Monarchs – who claim to represent your organism and rule over your mind, but who are actually controlled by a deeper, hidden system.

These Monarchs may be delusional, but they still seem like an important feature of our mind. What does science say about our self-awareness – our ability to survey our mental processes and recognize our own identity? And at what point in our evolution did we develop this capacity?

One experiment commonly used to test for self-awareness in animals is the mirror recognition test, first devised in 1970 by Gordon Gallup, Jr.[25] First, the animal is familiarized with a mirror. Then researchers place a mark on a part of the animal's body it cannot see directly, such as its forehead. The animal is then reintroduced to the mirror. If the subject touches or inspects the mark, or spends considerable time examining it, researchers interpret this as evidence of self-awareness. Great apes, elephants, dolphins, and some birds have successfully passed this test.

I must confess I am puzzled why this experiment has been

given as much credence as it has. It seems like a test of whether an animal can read, interpret, and remember visual information in a mirror, rather than a test of self-awareness.

Imagine you're set to leave your apartment to go to work – you've got everything you need: your wallet, your keys, your shoulder bag. As you are about to leave your home and pass by the full-size mirror next to the door, positioned strategically to catch fashion *faux pas* before they are exposed to the world, you see a weird stain on that bag slung over your shoulder. What is that? Will it come off? You try to rub the mark away.

Does this action prove you think your shoulder bag is you? No, it just means you saw a stain and tried to rub it out. The same logic applies to the animals in the mirror test – recognizing a mark and responding to it doesn't necessarily equate to self-awareness.

What animals have self-awareness, then? The set is probably far broader than that listed above.

The well-studied behavior of wolves demonstrates that they recognize each other through scent, appearance, and vocalizations. Their recognition of each other allows them to adopt different roles in coordinated hunting strategies such as scouting, driving, and finishing their prey (sometimes even by ambush, as our fireside storyteller proposed in the first chapter.) Wolves also share responsibilities like caring for pups and defending territory.[26]

To perform these roles, wolves must process both their own identity and the identities of their packmates, acting in alignment with their assigned social roles. Over time, as their role in the pack changes, their behavior adapts. For instance, if the alpha male or female – typically the only breeding pair in the pack – is killed, another wolf will step into the alpha role.

In computational terms, somewhere in their wolf brain

there is a property that represents their organism, and this property is assigned values: in the case of the wolf ascending to alpha status in the pack, the property assigned to his or her self changes. It's unclear how the behavioral operations they perform could happen otherwise. Wolves must have a sense of self.

You might object that this behavior does not prove that wolves have self-awareness; it might just be instinctual, unconscious behavior. But what do we mean by self-awareness? If the claim is that wolves process their identity when making behavioral choices but don't do so consciously, the distinction hinges on what philosophers of consciousness call *qualia*: the subjective experience of what something feels like. Since it is improbable that we will ever teach animals to articulate their qualia, it's unlikely we'll access their subjective experience.

If self-identity is a property that is processed, isn't that both as much as we are likely to discover but also as much as is useful to know? After all, as we saw with the Deluded Monarch of the last chapter, a description of qualia will not necessarily be a trustworthy report of what is going on underneath the hood.

Given the inadequacy of our insights into animals, the most we can say is that something like Self probably evolved as a necessary adjunct to complex social relationships in many animals. However, if we bring our inquiry to humans – whom we can, in experimental settings, instruct and debrief using language – we discover a richer trove of findings.

What do we know about how humans develop self-awareness? At the dawn of the 21st century, a serendipitous scientific discovery opened a window into how we develop our sense of self.

The Mental See-Saw

Marcus Raichle and his colleagues at Washington University in St. Louis were pioneers in a new field of research: using the new tool of fMRI to map mental functions to specific areas of the brain. This new capability allowed scientists to learn about the brain without having to wait for surgical mishaps or unfortunate maladies like Louis Leborgne's aphasia.

Their experiment aimed to have subjects perform a mental chore so that the researchers could map the results and see what parts of the brain were active during particular tasks. But as sometimes happens, the moment of epiphany came outside the boundaries of the experiment. The fMRI machine was capturing signals of activity even before the subjects had begun their assigned tasks. Many researchers before them had seen these sorts of images before a study and neglected or dismissed them. If they were conscientious, sometimes they would subtract the noise they saw from the actual results, since it was just noise.

Raichle and his colleagues began their experiment with just that intention. In the same way you would zero out the weight of the bowl before measuring the weight of flour for a cake, they were looking for the baseline of their measurement. But as they looked for that baseline, they noticed something strange. In subject after subject, there was a locational similarity in brain activity during idle moments. It was like there was a neural network for doing nothing.

As soon as the subjects started on a goal-oriented task, this network receded. Once the task stopped, however, the network returned. It was as if parts of the brain were on a see-saw: you could tip over into the goal-oriented task and activate some other network, but if you didn't you would tip right back into the doing-nothing network.

In 2001, they published a paper describing the Default

Mode Network (DMN),[27] and their findings led to a deluge of research. Since its discovery, there have been tens of thousands of mentions of the DMN in scholarly literature. You yourself know the DMN's work intimately, because it fills your idle moments – nearly all of them – quite fully.

The Idle Mind's Output

What do you think about when you think about nothing in particular?

You do some social rehearsal: you imagine yourself delivering a droll line at after-work drinks. Everybody laughs.

You think about your biography: it's funny how you always pick partners that never pick up after themselves. Why is that?

You do some grudge-nursing: why does Jeff at work always slink out of doing that task he hates and leaves it to you instead?

You do some planning: you will remember to get milk on the way home tonight, right?

You do some self-appreciation: one good thing about you is that you're always there for your friends... like with LuAnn last night. How did you manage to stand her complaints?

You do some self-criticism: why are you always so tongue-tied around Jay?

You worry: they said there's a 40% chance of rain on Saturday – it's going to ruin your barbeque and everybody is going to have to be indoors, with the AC still broken, and they will be miserable.

You do some dreaming: that promotion is more likely than not, and with the additional money... why not get a bigger, more comfortable place?

Broadly speaking, we can divide the content of our idle moments into three categories: a) Review, where we look back at our episodic memories, b) Anticipation, where we look forward to future events, and c) Self-definition, where we take stock of who we are.

REVIEW	ANTICIPATION	SELF-DEFINITION
Practical review of functioning/learning	Social rehearsal and strategizing	Self-evaluation of traits and abilities
	Practical planning and problem-solving	Positive (pride)
Biographical narrative-shaping		Negative (shame)
	Aspirational dreaming	Moral self-evaluation
Emotional processing		Positive (righteousness, anger)
	Anticipatory emotions (excitement, anxiety)	Negative (guilt)
		Evaluation of social circle, alliances, and enmities

In the last chapter we talked about the Deluded Monarch who imagines himself as governing our organism. It is in these idle moments when we imbue the Monarch with narrative and personality.

It seems like we are doing nothing, but during our daily mind-wandering, we are really quite busy: scanning our past life for a back story; planning our moves for the next act; and trying to figure out what our essential nature is. These thoughts might be small-bore: *don't forget to pay the electric bill!* Or they may be epic and consequential: *I need to quit my job and change careers.* But whether the stakes are small or large, the DMN is the storifying part of your brain. It is a bard that is constantly at work, writing and rewriting the character at the center of the play of your life: you.

The human brain is enormously energy-intensive. About

20% of the calories you consume power the brain, and it is estimated that 60-80% of the brain's energy goes to sustaining the Default Mode Network.[28] If the DMN provided no fitness-to-survive advantage, and if there were a version of humans with no DMN, then they would have an enormous advantage over us because they would require less food to survive. Our organism makes a huge investment in the DMN, so it stands to reason that it plays an important role in our organism's well-being.

What is the utility of this self-authoring mechanism?

The Self As Bridge

The construct of the Self has both a social function and an internal, regulatory function.

Human societies are highly complex, so (just like the wolves) we need to develop a concept of what our niche roles are – how we fit into the larger whole. This is true for our productive function, certainly: how our work contributes to others. Are we leaders or followers? Are we good with people, good with our hands, or good with numbers? Are we caring parents, or just doting aunts and uncles?

A concept of Self is useful for our social functions too: are we the empathetic listener? The connector of people? The clown that leavens gatherings with merriment? Our Self is always considering where it fits in and what the opportunities to maximize its prestige and relatedness are.

Once we have crafted our Self we strive to advertise its best qualities. By words and deeds, we let others know that we are reliable contributors to the communal entity, whether it be in social, familial, or productive functions. The Self is not just about who we are, but who we are to others – as parents, siblings, and children; as romantic partners and as friends; as bosses and as employees; as neighbors and as citizens.

The Self is a bridge between our organism and society, but bridges have two sides. On one side, the Self is looking for a fit. But on the other, it is shaping our behavior by creating and amplifying emotional signals.

Imagine you are working late to meet a deadline. What is going through your head?

Karen missed the deadline on her report last week, but I'm not like Karen. When I make a commitment, I keep it, damnit. I'm not going to leave the office until the report is done – even if it takes till midnight. God, remember last month when I used the wrong numbers for the report? It was so humiliating. The next day, everybody had to drop what they were doing and help me catch up. That's not going to happen this time. It's just a bit after 11:00 and I am nearly done. That's who I am: I work hard, and I am reliable. Tomorrow, the boss is going to notice that, and everybody on the team is going to notice that.

Your sense of Self is acting as an enforcer on your behavior. It is not enough to have a sense of who you are and let people know about it; you also need to deliver on that promise. When you live up to your concept of your Self, you feel the pride of a solid contributor. When you don't, you feel a torturous shame. You have let yourself down, and you have let the team down.

Sometimes, however, there is no other victim but us. When we have been let go of a job, when we have been dropped by a romantic partner, or when somebody we love has died, then the narratives woven into our Self will be cut short, and we will need to grieve for the dreams we have lost. It is in these moments, when there is a breach opened between our Self and reality, that humans feel some of our most agonizing psychic pain. These are times when we see ourselves as lonely, losing, failed, or fake – inadequate to the daily challenge of life.

The Self is a disciplinarian that stings us with an electric shock when we are not fulfilling our own ideal of ourselves, but sometimes we feel the sting even when the situation is beyond our control.

The ancient Greeks hypothesized that matter was made up of indivisible units called atoms. Whether our tendency is cultural or biological or a little of both, we tend to see our Self as Atomic – let's call it Atomist, so it sounds a bit less radioactive. The Atomist Self can change and grow, sure – but we view it as an indivisible, enduring line in our life. Every day we foster it and protect it, as if it was a cherished heart-stone.

Three Imperatives

As organisms, we have many needs; among them: food, shelter, and safety. These physical requirements are encoded in simple ways: a yearning for a hamburger, the feeling of cold blowing through you, the fear of a ledge.

The Atomist Self's work is different, more complex, and focused on our social interactions. To do its work as a social bridge, the Atomist Self has three main social imperatives it seeks to maximize: relatedness, structure, and prestige.[29]

The most devastating outcome for the human organism is ostracism from the tribe. Once are are ostracized our reproductive prospects are effectively zeroed out, but more than that, in our ancestral environment our likelihood of survival would be radically diminished. Because of this, the

Atomist Self's most foundational response to the world is the need for relatedness. To feel whole, we need to create bonds with our partners, our family, our friends, our tribe. If we do not have these bonds, the agony of loneliness will consume our lives. In many prisons, solitary confinement is the harshest punishment available. Atul Gawande has written movingly of how this method of social control does long-term damage to the mental health of prisoners and might constitute torture.

Almost as cruelly, the ritual of ostracism is acted out frequently in high schools. While boys can be violent bullies, the preferred form of sadism among some high school girls is to cast friends out of their cliques. The psychological results can be devastating.

Although friendships and group affiliations are important, the most significant relationship we are likely to have with someone who isn't a blood relative is an intimate relationship with a romantic partner. For many, this is the gateway to family life. Even those who choose not to have a family can find this bond deeply meaningful. From the point of view of our genes, the most important evolutionary imperative is reproduction, and that cannot be done without a partner. Although many people can forge happy and meaningful lives without a romantic partner, for others, lacking a romantic partner is a painful and stigmatizing burden.

The second imperative for the Atomist Self is structure. The tribes we are part of have their rules and processes, and each member of the tribe shares responsibility for enforcing this structure.

Perhaps the most primordial aspect of social structure is the desire for fairness. This need is probably hard-wired into us, since it seems to be existent in our ape cousins. You can find some very funny footage online of Frans de Waal demonstrating an aversion to inequity in capuchin monkeys.[30] Two monkeys sit next to each other in separate cages, both with levers to pull. When the first one pulls his lever, he gets a grape: a very desirable reward. When the second one pulls the lever, he gets a cucumber: a less desirable reward. The cucumber monkey puts up with this for a while, but when he has had enough he tosses his cucumber at the unfair experimenter in a fit of pique. Equal pay for all, he seems to be screeching.

We don't just demand fairness in our dealings from others; we demand it from ourselves. If we are given a gift and do not reciprocate, we often feel a sharp pang of discomfort. In our ancestral environment, where food-sharing was vital for survival, this must have been a useful norm that became reinforced by acquired instinct. Today, this impulse is taken advantage of by marketers and salespeople using free samples, lavish personal attention, free meals, and other offerings to extract from us a feeling that we are in the merchant's debt.

One of the most important elements of structure is that we give preference to those in our in-group. In our ancestral environment, tribal loyalty must have been necessary for survival. We still retain the need to be part of a group, but in today's world those groups have splintered from one tribe into many. Of course, our family grouping is a tribe, but we also have many other affiliations. The potential is in us to bring a tribal fervor to our workplace, our race, our religion, our nationality, our political affiliation, or even a sports team. While good things can come from such feelings of affiliation, the attendant disrespect and distrust to out-groups can be so furious that it spills into violence.

Structure does more than protect us from each other; it can also protect us from the encroaching tribe: needs such as privacy and autonomy can be considered to be part of our desire for structure.

The third imperative is prestige. This need is often referred to as *status*, but I've preferred to use the word *prestige* because it does not contain the connotation of bucking for higher socio-economic position. That is assuredly an important part of prestige, but I also want to include interpersonal needs where socio-economic positioning is not at all in play.

Despite our innate need for fairness, in our tribes everyone is not equal; hierarchies exist. In the workplace, these hierarchies tend to be defined precisely in organizational charts. In our social life, hierarchies tend to be defined in a more informal and blurry way. Of course, club memberships or the neighborhoods we live in can be clear indicators of class, but the game of status does not stop at a display of income. Our status can also be defined by the number and nature of social contacts we have. In this ambit, there are many levers for social success other than wealth and power: physical beauty, charm, intelligence or just good-heartedness can win you friends and connections. Yet – once again – these contours fall short of describing the totality of our very basic need for status.

In most of our social interactions – even the most fleeting and informal – an antenna goes up, telling us whether we are being valued, or whether we are being disrespected.

Just the other night I was returning home in a cab to my apartment in the beautiful city where I live, Buenos Aires. I was riding with a friend, and we were chatting in English. When we got to her stop, she got out, closed the cab door, and waved goodbye. The cab took off with me still inside. I hadn't registered anything unusual, but the cab driver started grumbling at this point. A couple of blocks later we got to the final stop, and I started looking in my wallet for the bills to pay the fare. The cab driver looked exasperated and said something like: "First you slam the door, now you're looking for exact change. Do you know how little that is in your money?" I was taken aback. First, because tipping is not expected in cabs here. But also, because the denizens of Buenos Aires are overwhelmingly *amables*. It's a very gentle and polite city compared to its brethren of a similar size. I rounded off the fare to pay him, and he looked me in the eye with a bit of a smirk and said: "Now can you please not slam the door?"

What is interesting me is not the incivility. That happens – anywhere in the world. Moreover, I cannot say what the cab driver was thinking or feeling, or the life context of his behavior. Perhaps he was highly sensitive to noise. Perhaps his door had been damaged before. Or perhaps he had suffered a bad night with few fares and he resented that the passenger in his cab probably earned more than him with less work. I cannot tell, so speculation is fruitless.

What I can examine was my reaction to it... and it was epic. The minute I got out of the car I started composing bitter responses to him, grinding and polishing the ripostes for a sharper edge, something that would have cut him back down to size. The experience cast a pall over the rest of my night – I was offended, hurt. What really had happened to cause this over-sized response in me? Yes, he must have imagined me as a caricature of the wealthy foreigner visiting his city, but what was that to me? He hadn't taken anything from me that I wouldn't have given him anyway. He didn't hurt or even inconvenience me. I might never see him again.

At that stage my existence had devolved into a battle of prestige. I had been relegated to lower place of rude, clumsy, foolish, stingy – definitions that did not accord with how my Atomist Self saw itself. I felt the need to fight back, even if it was after the fact and only in my imagination. For the cab driver, it must have seemed like he needed to fight back too. But chances are he said what he said to recapture some sense of lost respect, some prestige he felt life had unfairly withheld from him.

Why are the taxi driver and I fighting this battle? To defend a construct in our minds that is cultural but also psychological: our prestige among others.

When a workmate receives an assignment we would have liked to receive...

When somebody talking to us assumes that we are ignorant about a matter...

When we hear about a social event we have not been invited to ...

When our romantic partner doesn't seem to value our contributions to a relationship...

When a parent doesn't acknowledge our achievements...

When a friend says something disparaging about our skill at something...

Think back to last argument you had, or the last time your feelings were hurt. Chances are your prestige was at stake.

Winning the respect of those around us is a vital imperative of our Atomist Self. We have a story we tell about ourselves. We need to protect that story from others and, sadly, sometimes we feel the need to diminish the narratives of others to make our own shine more brightly by comparison.

Another way of describing the need for prestige is pride; it is a less specific label than prestige, but the mythic flavor is appropriate since it is hard to overstate the force of this urge. The Greeks considered pride the hallmark of hubris, the flaw that led to the fall of their tragic heroes. In both Christian theology and Milton's *Paradise Lost*, pride was the original sin that led Satan to be cast out of heaven. For many traditions, pride is the root of all sins.

The need for prestige – our insistent pride – is the origin for great achievements, but it is also the setting for a torturous psychic battle. That's because our need to notch a higher prestige in our social affairs is in conflict with our need for relatedness, and to a lesser extent, our need for structure. Our Atomist Self contains a deep, warring contradiction within itself.

Taming Monkey Mind

Like birds making nests out of twigs they have fetched, we make our Atomist Self out of thousands of stories we hear and adopt as our own. Our culture and biology does not make it easy to see the reality of a modular, impermanent, and interdependent mind because there is a biological and social function to the Atomist Self. This Self disciplines a social strategy that helps us navigate society.

At the same time, the Atomist Self is also the source of our profoundest pains and dissatisfactions. If the Atomist Self did not hurt us, it could not do its job. It needs to apply the sting to keep us heading where we're supposed to go.

The chattering of the DMN that creates our Atomist Self also creates what the Buddha called *upadana*, meaning clinging or attachment. These are the self-definitions our self-talk provides for us during our waking life. The ensuing discomfort and unease that *upadana* causes is what the Buddha called *dukkha*.

This endless geyser of self-talk that accompanies our idle moments is never going to be stopped all together, nor is stopping it entirely something that we should want. But there is solid evidence that practice can gentle the gush.

Several fMRI studies have shown that meditation reduces

DMN activity.[31] Moreover, this effect lasts beyond the period of meditation itself; the baseline of DMN activity for meditators is lower than it is for control groups, suggesting a lasting impact on Atomist-Self-authoring.

Is there evidence that meditation affects emotional regulation and creates positive affect, which is to say, happiness? Again, there have been longitudinal studies showing impressive gains after even modest meditation training, on a wide range of measurement tools.[32]

In one exceptional case, neuroscientists at the University of Wisconsin–Madison used EEG sensors to capture the brainwave activity of French-born Tibetan monk, Matthieu Ricard. What they saw during his meditation was a surge in gamma waves – often associated with positive affect – greater than anything their tools had captured before. This led the Smithsonian magazine to dub Ricard "the world's happiest man."[33]

There is even proof to bolster a bolder claim for meditation: that it fosters compassion. In a well-known experiment, David DeSteno and his colleagues at Northeastern University randomly assigned participants to two groups: one given an 8-week meditation course, and another that was not. After the training, participants from both groups were invited to an office, unaware that the real experiment was taking place in the waiting room. There, three chairs were arranged, with two already filled by confederates of the researchers. The subjects entered one by one, at their appointed times, and naturally took the third chair that was still free. Then a third confederate entered the waiting room, assisted by crutches and wearing a surgical boot to simulate injury. Fifty percent of the meditators gave up their chair, but only 16% of non-meditators did, suggesting that meditation fosters not just happiness but also empathy.[34]

You would think that, with all this evidence, loosening the hold of story on us would be the dominant form of treating psychic pain in our society – and of course there is no shortage of advocacy on behalf of mindfulness methods. However, the dominant strand of thought in our society is exactly the opposite: it argues that the way to treat pain is just to write a better, more flattering story of our Self. This is the philosophy of self-esteem.

This school of thought tells us we should be on guard against negative thoughts and replace them with optimistic views of our capabilities. We should praise our children and give them trophies for participation. We should create safe spaces that shield students from criticism and questioning. Books like *The Six Pillars of Self-Esteem* and celebrities like Oprah Winfrey remind us that the way to a more fulfilling life is by valuing our Self more – leaning into story.

The problem with this methodology – and the reason why for most it is doomed to fail – is that it just heightens cognitive dissonance. Even if you successfully managed to will yourself to a more positive view of your powers, and this allowed you to approach the next challenge more confidently and successfully, eventually a shortcoming of some sort is sure to appear. When that happens, if the gap between your self-image and reality has widened, then your despair will increase too. Psychologist Edward Tory Higgins called this breach a self-discrepancy. All narcissisms – even mild ones – are necessarily fragile.

The only way to reliably mitigate the pain of story is to loosen its grip on us.

The Empty Boat

Remembering the story of my encounter with the cabbie, I was reminded of the story of the empty boat. The way it is usually told is something like this:

On a beautiful day, a man seeks some quiet by taking his skiff out to the middle of a quiet river, where he lays anchor. He rests in the sun, enjoying his natural surroundings.

Not long after, a boat approaches, heading toward his own. The man spots it coming and yells to caution. No one responds.

Now concerned, he yells again. Still nothing. And then, crash! The boat has collided with his own.

Now the man is furious. But as he inspects the damage, and peers into the offending craft, he realizes it is empty. The boat must have been pulled off its moorings and been carried along the water by the wind. He has no one to be angry at.

My encounter with the cabbie was like that, wasn't it? The trespass was empty of any real harm. The man was just a passing figure of no lasting weight in my life. I was getting angry at an empty boat. This is especially true since most of my anger occurred after the event was over, in the fervor of my vengeful imagination. I was truly swinging at a phantom.

That is a good lesson, but if you go to the original text by Taoist philosopher Chuang Tzu (4th century BC) there's a different method and a deeper moral. Chuang Tzu is less dramatic, but also subtler. He switches the point of view to that of the empty boat:

If you can empty your own boat
Crossing the river of the world,
No one will oppose you,
No one will seek to harm you.[35]

The solution is not to fortify the periphery of your boat, as self-esteem advocates would have you do.

Rather, it is to loosen the Atomist Self's hands from the helm, so that you might become an empty boat too. If you can do that, then it is unlikely that you will be offended by other boats' plaintive pleas for prestige in this world. And they will be less likely to be offended by you.

CHAPTER 4

CARROTS AND STICKS

T hus far in our evolutionary saga, the storyteller-to-be has emerged from the primordial pond and now has a Self – not just a consciousness, but a consciousness aware of its consciousness. Among the horde of processes teeming in our minds, one stands out as uniquely intertwined with storytelling: our ability to experience and evoke emotions.

When we lean back on the couch to see a movie at night, we are often hoping for an emotional experience – laughter at incongruencies, sorrow over injustices, or the gratification of love found. Likewise, when we lean forward after dinner to tell a story over a glass of wine, we would like to stir emotions too: the warmth of companionship, the resonance of shared experience.

Storytelling marketers – and for more than a decade I have made a career as one – have long known that while logic is invaluable, it is emotion that captures hearts and eventually closes the deal.

Likewise, in the story that opens this book, our ancestor Shiki conjures emotion so that he might persuade his tribe to act together. First he flatters them, then he evokes feelings of tribal affiliation.

Given that our capacity to feel emotions is fundamental to our ability to hear and tell stories, how – and why – did emotions first arise in animals?

The first Evolutionary Biologist was also the first Evolutionary Psychologist. After *The Origin of the Species* and *The Descent of Man*, Darwin's third major tome was *The Expression of the Emotions in Man and Animal*. In it, he theorized that emotions were an adaptive response that increased an animal's chances of survival. For example, an expression of aggression might deter threats, or an expression of affection might strengthen social bonds. The survival benefit of emotions wasn't purely communicative, however. He described how emotional responses might prepare an animal to respond to threats; for example, fear might increase the heart rate, allowing the animal to better flee from a predator. And Darwin also allowed for the possibility that emotional responses were vestigial of what might have been practical responses earlier in the evolutionary process. For example, expressions of anger often involve the baring of teeth, a trace of their original purpose as a threat display.

Later that same century, the intuitive view of emotions – that they are felt, and they then trigger their associated physical responses – was challenged by William James in his 1884 essay, *What is Emotion?* James' theorizing was duplicated independently by Carl Lange in Denmark, so we refer to their idea as the James-Lange theory.

Their view is that we don't run away from a bear because we are afraid; we are afraid because we are running away from the bear. To be clear, the physical reaction is not properly speaking antecedent to the emotional reaction; it *is* the emotional reaction. When we are under threat, our heart rate and breathing quickens, our pupils dilate, our skin perspires, our muscles tense – all these responses are meant to aid a fight-or-flight response. This physiological activity is not an effect of fear, it *is* the feeling of fear. This insight recognized something that has been verified by scientists repeatedly since then: that the body and the mind are inextricably linked and that much of what we think of as mind matter is happening because of changes in parts of the body outside of the brain.

One might expect that the century following James would have been dedicated to empirically testing these and other theories of emotion, but instead emotion as a topic of research fell decidedly out of fashion. The ascendent school of behaviorism thought that since we could not observe internal states, psychologists should only pay attention to what they could reliably observe, which was behavior. From Ivan Pavlov to B. F. Skinner, these theorists clarified many details about how we learn, but by the 1970s their avoidance of what was intuitively obvious – that emotions have a role in determining that behavior – might have partly contributed to the waning of their reputations. A new generation of scientists began to dig into the neglected topic of emotion.

One of the most important researchers in this newly invigorated field was the Portuguese-American neuroscientist Antonio Damasio. Damasio built on William James' insight about the physical nature of emotions, but he modified and elaborated on it in significant ways.

Elliot

In his influential book *Descartes' Error*, Damasio writes about how one day in the 1970s he was asked to see a patient he refers to as Elliot. This case history is a parable of how emotions do the work of the Atomist Self.

Before suffering the affliction that brought him to Damasio, Elliot had a wife, family, and an enviable amount of professional success when he started suffering from headaches. It became difficult for him to concentrate, and his work colleagues reported that his previously reliable sense of responsibility seemed to flag. His general practitioner suspected a brain tumor, and a specialist confirmed the diagnosis: Elliot had developed a benign growth called a meningioma. Even a benign growth can be dangerous, but this one was the size of an orange and growing rapidly. If not removed it would soon kill him.

Elliot was operated on and the tumor – which had been growing above the nasal cavities but just below the eye sockets, pressuring the frontal lobes upward – was removed. As is customary in these cases, a lot of the brain tissue that was damaged by the tumor was also removed.

After the operation, Elliot's cognitive faculties were tested, and it seemed like they were largely intact. It soon became apparent, however, that his personality had changed.

At work, the highly efficient operator of before now seemed to have a great deal of trouble prioritizing his time. Elliot seemed to lose sight of the overall purpose of his actions and focused his time excessively on minutiae. He would miss important meetings to finish off trivial tasks. His colleagues had to fill in for him so that his assignments might be finished. Eventually his poor decision-making caused him to lose his job.

Elliot's personal life suffered as well. His wife divorced him, and he entered a brief and ill-advised second marriage. He bet his life savings on a venture with a shady character, even after having been warned against it by informed advisors. The business ended in bankruptcy; Elliot's life was spiraling out of control.

When Damasio met him, Elliot was living with a sibling, unable to take care of his own needs. He had just been denied disability insurance, and Damasio had been asked to certify that he was not just a malingerer but was indeed suffering from a genuine malady. Damasio describes Elliot when he met him as charming, sophisticated, ironic – even comparing him to the Addison DeWitt character played by George Sanders in the film *All About Eve.*

Damasio subjected Elliot to a battery of cognitive skills tests – long- and short-term memory, problem-solving, language processing, etc. – in all of them, Elliot emerged as either average or superior. Perhaps the problem was with his personality, then? Elliot took a personality test, and there too he scored within the band of the normal. Damasio concluded:

> *After all these tests, Elliot emerged as a man with a normal intellect who was unable to decide properly, especially when the decision involved personal or social matters.*[36]

So what *was* the trouble with Elliot? From the first time he took his case history, Damasio noticed that Elliot told his story – with all its agonizing turns – with a flat affect. In fact, Damasio seemed to feel more anguish about Elliot's travails than Elliot did. This might have been dismissed as a patient who simply had a stoic disposition, but as Damasio got to know Elliot better he observed his emotional deficit seemed to go considerably beyond a stiff upper lip. When this theme was brought up, Elliot self-reported that, since his operation, topics and situations that once evoked his strong emotions now left him unmoved. Writes Damasio:

> *I began to think that the cold-bloodedness of Elliot's reasoning*

prevented him from assigning different values to different options, and made his decision-making landscape hopelessly flat.

Damasio tested this theory experimentally. Elliot, along with a cohort of other patients with frontal lobe damage and a control group of uninjured subjects, were shown a number of photographs – some of them quite bland, but some of them showing disturbing images of violence or harm. While the subjects viewed these images, their galvanic skin response (GSR)* was measured. As predicted, the control group's GSR peaked when they saw the disturbing photos, but the response of Elliot and the frontal lobe patients was relatively flat.

Very well, but could this emotional deficit be tied to decision-making? To test this, Damasio conceived of a gambling game experiment. The subjects were asked to draw cards from any of four decks. Each card in the decks would have rewards or penalties attached to it. Two of the decks reliably granted the player-subjects a small play-money reward, and two reliably provided a larger reward. The catch was that the two decks with the more sizeable rewards also had penalty cards that undid the benefit of the rewards. The decks with the smaller rewards had smaller penalties that did not completely reverse their benefits.

Elliot, like the rest of the frontal lobe patients, kept going back to the dangerous, high-reward decks instead of the lower-reward decks that were safer long-term. The subjects without cerebral lesions, on the other hand, eventually caught on to this

---

* The GSR measures the conductivity of skin, as it is affected by subtle and unintentional sweating. The test has a long history in psychological research as a reliable way to measure emotional responses. It is also part of a triad of measures that makes up the so-called lie detector test.

pattern and chose the lower-reward but safer decks. Despite having sound cognitive faculties, it seems that Elliot and the other injured patients lacked fear emotions to guide their decision-making.

The work of Daniel Kahneman and Amos Tversky has shown that generally we humans are irrationally averse to loss,[37] but Elliot's problem was the opposite: his ability to *feel* loss was so impaired that he was not able to make rational decisions to protect his gains. Emotions, in Elliot's case, were like an internal probability calculation that measured possible gains against possible losses. The control group in the card experiment acted depending on which deck had a better greed-to-fear ratio; Elliot and the rest of the frontal-lobe patients couldn't make this same calculation accurately.

To return to our Parliament of the Mind metaphor, we might think of emotions as a way to shift the will of the Parliament: a form of lobbying, a shout in the well, a hand raised to vote emphatically for an outcome. Emotions likely evolved to fulfill a communicative function as Darwin theorized, but also to help mediate between contradictory responses of the many modules of our mind. (*Which is stronger: my excitement about this deck of cards, or my fear of this deck of cards?*) Elliot's case study teaches us not that reason is subservient to emotion (although of course, experimental results that demonstrate motivated reasoning are legion). Nor does it prove that emotion is subservient to reason (as many poor investors – riled by irrational fear or exuberance – have proven for themselves). What it does show is that emotion is an important part of our decision-making process, and that the two faculties – emotional and rational – work together in subtle and important ways.

So, what is it that distinguishes emotion from reason? The physicalized-in-the-body nature of emotion is a major distinguishing factor. On the cognitive plane, another difference is how emotion can create narrative spirals. If my wife is worried about the safety of the brakes on my car, her anxiety might include a logical evaluation of the risks, but it will be dominated by the repeated mental screening of many scenarios: some leading to my demise, some leading to hers, and some leading to the dismemberment of innocent pedestrians and adorable pets. The emotion of these stories-on-repeat not only spurs her to action, but it also allows her to communicate the gravity of the situation to me in an effective way, so that I will not have a calm moment until the car has been (quite rightly) taken to the mechanic for inspection. A key aspect of emotions is their shareability, hence their importance in storytelling.

The Emotional Mechanism

If we are donkeys and the Atomist Self is our rider, then emotions are carrots placed in front of us to spur us toward boons, or sticks poised at our back sides to deter us from threats.

Fear, for instance, keeps us away from dangers.

Anxiety draws our energy and attention to potential problems.

Infatuation pulls us to a potential mate.

Guilt punishes us for violating the norms of our social group.

Pride rewards us for achievement that is prized by that group.

In all these cases, emotions are directing our attention to valuable information. They are steering us toward what is good for our survival and reproductive prospects, and away from what is bad for them. Sometimes they spur us into action; sometimes they implore us to focus on a matter; and sometimes they fine or reward us for what we've recently done.

So if emotions are looking out for evolutionary wellbeing, why do they feel like *dukkha,* or dissatisfaction?

The generation before mine had a song about not finding satisfaction. (It was sung by a fellow named Mick.) For my generation, the musical touchstone about not finding contentment was The Replacements' *Unsatisfied.* As sung by Paul Westerberg, it's a plaintive cry from a boy who became the heralded rock star he always wanted to be, but found the experience lacking:

> *Everything goes*
> *Well, anything goes all of the time*
> *Everything you dream of*
> *Is right in front of you*
> *And everything is a lie.*
> *Look me in the eye*
> *And tell me that I'm satisfied...*
> *I'm so unsatisfied.*

Why is it that even when we reach our dreams, it's still not enough?

Psychologists and evolutionary scientists have quite a bit to say about this. From an evolutionary standpoint, we're not meant to catch the carrot. If happiness were a constant, it would fail to motivate us. Full contentment is fleeting by design, keeping us hungry for more.

The good news is that the opposite is also true: our discontent doesn't last long either. An impressive body of research has shown that our affect tends to return to a baseline level, like a thermostat regulating temperature. One famous study interviewed lottery winners and victims of serious accidents that left them paralyzed.[38] While the lottery winners considered their win to be a positive event, and the victims considered their accident to be a negative event, in both cases their affect tended to return to their pre-event level. This phenomenon is known as the *hedonic treadmill*: we adapt to our circumstances, and their power to influence our happiness diminishes over time.

Moreover, we are wired to misjudge the workings of our own emotional system. Numerous studies have asked subjects to forecast how they will feel about positive or negative eventualities in their lives. In one experiment, college students were asked to forecast how the outcome of a lottery assigning them to either a desirable dorm or a not-so-desirable dorm would affect their future happiness. They predicted a significant difference, but in actuality, follow-up assessments revealed that students in both dorms reported nearly identical emotional well-being.[39] (Tangentially, I can attest that my life was not *completely* ruined by being assigned to NYU's Weinstein dorm.)

This tendency to overestimate the emotional impact of future events makes sense from an evolutionary perspective: if emotions are designed to motivate behavior, they work best when we believe their impact will be profound – even if their actual effect is fleeting. At the same time, it is adaptive for emotions to quickly return to a baseline level, so they don't interfere with ongoing action. This rapid return also maximizes the perceived gap between present and future happiness or sadness, reinforcing motivation.

Over the long term, our emotions regulate to a baseline. In the short term, however, the Atomist Self fools us into thinking that they will last. And since the Atomist Self sees itself as permanent and central, it often ascribes the cause of these emotions to its own immutable qualities; it interprets fleeting emotional weather as a lasting climate.

For instance: if we score low on a test, we can feel like we will always lack the intellectual gifts to succeed at a task, ignoring how better preparation might make a difference.

If we are victimized by crime, we can feel like we are personally weak and will always be in danger.

If a potential mate rejects us, we can feel like we are inadequate for every mate we might desire.

If we recognize that we have wronged someone, we can feel like we are morally deficient and perennially guilty.

Emotions are transient phenomena that happen to our organism. However, when we take these valuable signals and etch them into a narrative that represents our soul, our worth, and the arc of our life, we can turn a pebble in our pocket into a boulder on our shoulders. Temporary setbacks and obstacles can assume existential scale.

We have referred to the stories that make up our Atomist Self as *upadana* (meaning clinging or attachment) and the emotions that arise from these stories – guilt, shame, anxiety, and others – as *dukkha* (meaning dissatisfaction or unease.) We have said these are adaptive responses. If *dukkha* and *upadana* are created by evolution, doesn't that make them natural? And isn't what is natural good?

When we go shopping for food, we perk up when we see the label "All Natural." When Aretha sings "You make me feel like a natural woman," we gather she means she is feeling good. When we go to the doctor and note a minor symptom, the doctor waves off concern with an airy "that's natural for someone in your situation." Nature is the norm, the benign default. It is the grand force that is looking out for us.

But is it?

Viruses are natural too. Hurricanes and tornadoes are natural. My cat Hildy torturing a mouse she has caught – not to eat, but simply for her amusement – is natural. If whatever comes from nature is natural, then the impulse to steal is natural.

The notion that what is *natural* is necessarily "what is good and right" is what philosophers call the naturalistic fallacy. The tools and impulses that nature gives us are not always socially constructive or acceptable – the behavior of the faithless cad who sires children by many women without paying heed to their upbringing, or the aggressive bully who enforces his will with violence – these might all be understood from an evolutionary point of view, but that doesn't make the behavior right. The ethical schemes of human culture are independent of, and very different from, natural laws.

Perhaps you have a different reservation. It's not that you think nature should be accepted as it is for ethical reasons, but you suspect that there are practical benefits to *dukkha*. Perhaps you identify with Nietzche's aphorism:

Be careful when you cast out your demons that you don't throw away the best of yourself.

Some people are a little in love with their demons. They feel they need an inner taskmaster's insults and badgering to discipline their life quest. The last thing they want is to quiet the Atomist Self's constant thirst for prestige because… they *want* that prestige! They are okay with living in that self-reinforcing loop. We need a name for this attitude so we can refer back to it, so let's call it Molly Strivers' credo. (Apologies to anyone actually named Molly Striver.)

One of the unremarked dignities of the human condition is that your relationship to yourself is largely private. If you feel like you need the lashings of *dukkha* to be your best, then tyrants and saints are powerless to intrude on your inner preserve. Molly Striver is free to choose the life she wants – remarkably free. And so are you.

However…

You should not assume that an attachment-reducing practice leads to a lower social or productive functionality.

The Cost Of Story

The pain the Atomist Self creates is a cost. When we look outside ourselves, we are trained to consider the causing of pain as unethical by definition. But when Molly Striver chooses pain for herself, chances are we don't see it as evil or wrong – we just think that Molly is being a bit hard on herself. When it comes to self-torture, we are ethically quite lenient.

Nevertheless, is this pain giving Molly a practical leg up on others not thus disciplined? We measure this *not* on the ethical scale we are talking about, but on Molly's (and her Atomist Self's) criteria of prestige? The answer probably depends on the person. You might know super-achievers who are driven by insatiable egos to ever-higher career rungs. We can't pretend that such people don't exist, or that their inner stories aren't what's driving this forward charge.

Still, that's a bit mealy-mouthed for Molly's taste. "If the Atomist Self didn't give me a fitness advantage, nature wouldn't have given me one," she says.

Well, Molly Striver would be correct, with some contextual provisos.

Yes, the Atomist Self undoubtedly provided a fitness advantage...

1. on a population average,

2. in our ancestral environment, and

3. on net.

Let's take a closer look at these conditioning phrases, starting with *on a population average*.

The Atomist Self may be beneficial to the population as a whole, but it can be harmful to certain segments of that population. Genetics offers many examples of such traits. For instance, the gene for sickle cell anemia provides protection against malaria, which explains its prevalence among the populations of sub-Saharan Africa. However, when an individual inherits the gene from both parents, they develop sickle cell anemia – a painful and debilitating genetic disorder.

Similarly, the "cliff edge" hypothesis in genetics suggests that a trait may enhance reproductive success... but only up to a certain point. Beyond that line of precipice, the same trait can become a detriment. Creativity, for example, can lead to success and recognition, but in excess, it may contribute to mental illness. The long list of great artists who have struggled with their mental health – spanning from Robert Schumann to Sylvia Plath – supports this notion.

So yes, the Atomist Self provides splendid career results for many dream-driven climbers. It's also true that this orientation fails miserably for many others. There is a name for the self-loathing that spews out when we don't live up to the narratives our Atomist Self has created: it's called depression. According to the World Health Organization, 5% of adults worldwide suffer from it.[40] In the U.S. suicide is either the second- or the third-leading cause of death among people aged 10–34,[41] depending on whether homicide is having a banger year or not. For many of these people, the stories they tell themselves are literally killing them.

Of course, you don't need to suffer from full-blown depression to be dragged down by the emotions the Atomist Self uses to achieve its ends. For instance, the anxiety that can be productive by bringing our attention to important issues can also be destructive. It can tangle up our cognitive processes, distorting, delaying, or paralyzing our decision-making. And it can cause us to project a lack of confidence that makes our associates lose faith in our ability to perform.

Worse still, the Atomist Self's sting can cause us to retreat from challenges all together. For many people, performing below a level of perfectionism is inconceivable and deeply painful, so they choose not to risk failure. This, in turn, makes it difficult for them to develop skills that are comparable to their innate capabilities. Many of us know brilliant, profoundly talented people who have chosen not to engage the commons with their abilities – perhaps because they fear that the response will not match their pre-conception of what they should receive, or because they don't trust themselves to withstand rejection.

We've talked about the Atomist Self being caught in loops of negative emotionality, but when the Atomist Self seeks positive emotionality, the results are not necessarily positive either. When we need the inhibition-lowering effects of alcohol to feel confident, or the stimulatory effects of cocaine to feel powerful, we are going down roads that could be detrimental to our life prospects.

It is not just substances that cause destructive positive-emotionality loops. The gambler that needs a payout to feel like a victor, or the cad that needs the thrill of an illicit romantic conquest to feel emotionally whole; both are indulging in behavior that is potentially devastating to their lives.

Let's now move to the second contextual provision to the notion that the Atomist Self is an unalloyed boon to our fitness: *in our ancestral environment.*

An adaptation that made us fitter for our ancestral environment does not necessarily make us fitter for our modern environment. A taste for the sweet served our forebears by steering them towards high-calorie fruits that provided fast energy. Any excess consumption was stored in fat that had the potential to help us survive cruel winters. Today, however, when sugar has been refined to a purity unknown in prehistory, and corporations have learned how to tickle our instinctual buttons, a taste for the sweet is a recipe for obesity and disease. We can think of this sort of adaptation as anachronistic: something that worked in our ancestral environment but doesn't work in our modern environment.

Likewise, our ancestors lived in small tribes, with a limited number of social connections. Their world was very different from our accelerated, hyper-individualized, and hyper-networked environment. Just as how we did not evolve to deal with the potency of sugar, we did not evolve to deal with the potency of social media signals. Today, we face a society-wide addiction that fuels anxiety and saps productivity on a massive scale. Social media takes advantage of the Atomist Self's need for social affirmation, turning us into mobile-phone junkies. Our insatiable thirst for Likes has led us down alleys of pointless political self-righteousness, pseudo-celebrity posing, and the purposeful pursuit of the envy of others. Social media is a Roman-Coliseum-like battle for prestige.

For children in particular, this kind of technological innovation has had ruinous results. Both the 2024 U.S. Surgeon General's report and Jonathan Haidt's *The Anxious Generation* have exhaustively documented how social media has harmed the mental health of our next generation.

Having made these points, let me acknowledge: it would be seriously overstating my case to say that the Atomist Self is a completely anachronistic adaptation: in many ways, it still serves us very well. Indeed, it would be difficult to function socially without an Atomist Self.

Finally, let's address the third and last provision: *on net.*

When a random genetic mutation introduces a new feature to an organism, it might increase fitness, but it might also introduce a new liability. It might be a plus and a minus, but a plus *on net.* The long neck of giraffes, for instance, gives them access to higher foliage as a nourishment source, but it also creates a vulnerability to predators when they bend down to drink water, and it creates the need for an astonishing cardio-vascular system to get blood to their head. Evolutionary theorists sometimes refer to these as *trade-off adaptations.*

Sometimes we talk about evolution as though it were *brilliant design* – and it often seems like it – but it's important to remember evolutionary changes arrive by testing random variations. It is an unguided improvisation that can get twisty and hairy. A scene from the TV series *Louie,* with Louis CK and the incomparable Charles Grodin, captures the sense of this well. Louie stumbles into a doctor's office, in obvious pain. He tells the Doctor that his back hurts, but the Doctor say there's nothing he can do about it:

DOCTOR: The problem is you're using it wrong. The back isn't done evolving yet. You see, the spine is a row of vertebrae. It was designed to be horizontal and people came along and used it vertical. Wasn't meant for that. So the disks get all floppy, swollen, pop out left, pop out right. It's going to take another 20,000 years to get straight. Until then it's going to keep on hurting.

LOUIE: So that's it?

DOCTOR: It's an engineering design problem, it's a misallocation. We were given a clothesline, and we're using it as a flagpole.

LOUIE: So what should I do?

DOCTOR: Use your back as it was intended. Walk around on your hands and feet. Or accept the fact that your back is going to hurt sometimes. Be very grateful for the moments when it doesn't. Every second without back pain is a lucky second. String enough of those lucky seconds together and you have a lucky minute.

As an interpretation of natural history, the accuracy of the details may be wanting, but the insight is valid. Evolution is not an artist's finished masterpiece; it's a meandering, provisional sketch.

There is another example of a trade-off adaptation sitting by my feet right now. Like most felines, my cat Hildy spends about four hours a day grooming herself. There are fitness advantages to this: it helps control parasites, reduces odors that might attract predators, and helps regulate her temperature, among other benefits. But licking her fur so much also consumes a huge amount of energy, and the byproduct – hairballs – can, in rare cases, lodge in a cat's intestines and become life-threatening.

Human evolution is filled with similar trade-offs. Our "design" is full of them. An important one came early in our development as a species.

We have used the concept of an arrow as a metaphor for story and talked of what an enormous effect Story has had on human destiny. But beyond the metaphor, a sort of arrow – a spear – played a decisive role in our fate. Not long after we learned to hunt animals with pointy sticks, a problem arose. Most of the time, the wounded animal would escape. We would try to run the prey down, but since we still had the furry skin of an ape, we would become overheated during pursuit. In response, evolution led us to lose the fur and start sweating from our pores, creating a cooling system that made us extraordinary long-distance runners.* With our new nearly hairless skin and cooling system we were able to outlast many more wounded animals in chase than before.[42]

However, solving one problem created a new one. Newly naked, we were now exposed to the cold of the night. Our great ape relatives, the gorillas, have thick fur that keeps them snug in the cold nights of the African highlands. Our evolution into furless perspirers caused us to lose this form of heat regulation. It was a trade-off adaptation.

Luckily, ingenuity provided a solution and culture passed it on. The very prey we sought for food had hides, so we stripped these skins from their corpses, cured them in the sun, and put them over our bodies for warmth at night. In the day, we could shed the skins; at night, we could don them again. With this new approach, we could adapt on the fly to heat or cold.[43]

The Atomist Self is an adaptation such as this. Like furless skin, the Atomist Self improved our fitness. But like furless skin, it introduced a new burden: the pain and disconnection it needed to wreak to do its work. We needed the equivalent of a hide-for-warmth; a cultural adaptation that would ameliorate our new liability. We needed a practice that would temper the Atomist Self.

---

* There are only a few species that can match or better humans on long-distance endurance, horses and wolves among them.

The Connected Self

The solution we arrived at was something of a mind hack – but one of enough cross-cultural presence to suggest it has stowed away in our genes.

When we quiet the Atomist Self, another self can emerge: the Connected Self.

Mindfulness is one way to approach the Connected Self, but other spiritual traditions point the way to it too. I'll offer a more precise definition of the Connected Self in later chapters, and explore the research into its neural correlates. For now, I'll allow myself a more impressionistic description.

When we experience the Connected Self, we feel a profound sense of unity – with the world and with others.

The Connected Self is our original self. When we were newborns, fresh from the womb, there was no concept of *me* and *not-me*. There was hunger, there was fright, but there was no separation. Mother, baby, the blurry world beyond – it was all one. The Connected Self is a return to that mind.

No metaphor is quite adequate or complete to describe the Connected Self, so you need to pile metaphors on top of each other and blend their truths together.

The Atomist Self is a motor; the Connected Self is a radiator.

The Atomist Self peers through a distorting lens; the Connected Self removes the lens to reveal the world as it truly is.

The Atomist Self is a bridge to society; the Connected Self is a causeway to the mainland.

The Atomist Self battles; the Connected Self surrenders.

The Atomist Self is inner noise; the Connected Self is inner peace.

The Atomist Self seeks a fair exchange; the Connected Self gives freely, without keeping score.

This is not an on-off switch. Very few, if any of us, will

turn off the Atomist Self for good, plugging into a Connected Self as a new and different way of being. Rather, the moments of grace and hints of the numinous that the Connected Self brings will temper our Atomist Self, keeping it from ravaging our lives.

The Connected Self does not belong to any one culture or spiritual tradition. It is a species-wide trait. While it may be supported by neurological hardwiring, it has also been shaped by cultural practices – and the two have co-evolved, reinforcing one another over time.

All wisdom traditions tell a similar story: it's the tale of how a selfish, self-centered life is changed by an awakening, and how a new and luminous peace is discovered through service to others. This transformation is often framed as an act of surrender. In monotheistic traditions, this is the message of the Torah, the Christian Testaments, and the Qur'an. In Eastern religions, it is the teaching of the Bhagavad-Gita and the sermons of the Buddha. Even secular thinkers – practical philosophers like the Stoics and Confucius – arrive at the same conclusion.

By relieving us from the imperative of Atomist performance, the Connected Self lightens our mood. It brings us joy and meaning.

Ironically, the surrender is not much of a sacrifice in terms of prestige because while our society often raises monsters consumed by their egos to high places, it also exalts those possessed with the spirit of a Connected Self. Most of us would prefer teammates that put the team above their own interests. We celebrate the generous, the self-less, the other-directed. The Atomist Self knows this and often acts accordingly, but since score-keeping mode is still on there's no joy in the generosity.

In coming chapters, we will talk about the neurological basis for this state, and the different ways religious practices facilitate it.

The Donkey's Plight

For now, let's imagine the donkey hinted at in this chapter's title and imagine his predicament more vividly.

An ingenious but ruthless master named Darwin has strapped an infernal harness mechanism around the Donkey's body. This contraption features a compass mounted on top, connected to a stick dangling a carrot. If the compass needle veers toward the east, signaling that the donkey has strayed from a true-north path, the carrot swings westward. If the needle veers west, the carrot swings east.

As if that weren't enough, the donkey's legs are bound with hobbles linked to motion sensors. If the mechanism detects a lack of movement, a spring-loaded stick snaps out to deliver a stinging strike to the donkey's backside. The donkey will feel a smarting shot of pain. If the donkey wanders more than 20 degrees off due north, the stick is also fired.

The donkey is living a life of toil – a life lived with the anticipation of hurt. Will he ever catch the carrot? Will he ever be reprieved by the stick? Will he ever have a moment of rest?

What if that donkey could just be conscious of his hoofs clacking on the rocky path below him, conscious of the gentle wind as it curls up the mountainside, conscious of the rhythmic flap of his tail against his thighs – and in that moment of donkey consciousness, the infernal harness would disappear, vanish into the dry air? Strolling north would not be a forced march; it would just be what a donkey does.

That harness is our Atomist Self. Liberation from it is possible, for the donkey in the story – and for us.

CHAPTER 5

REMEMBRANCE

To tell a story is to remember, sure. But that makes the relationship between these two capacities sound simpler than it really is.

I have a memory from my childhood that I know can't possibly be accurate. It's fourth grade, I am living in Australia, and the teacher is giving a social studies class. She talks about how some populations in the warmer climes have olive-colored skin, and at that point I can remember quite clearly everyone in the class spinning around, pointing at me, and saying: "Yeah, like him!"

Now, I am quite sure at least one boy in the class must have said something like that, and maybe a couple of others turned to see what he was pointing at. But my memory has almost certainly exaggerated the rest – amplifying the response to match my discomfiture at being the center of attention. The memory became emblematic of a particular narrative of racial difference that, truthfully, wasn't that prominent in my childhood.

So yes, story requires memory, but also: memory requires story. Story is the compression-decompression algorithm that memory lives in. We will say more about the mutability of memory later in this chapter, but for now it's important to note that this human capacity is far from a monolith.

Memory comes in different forms. Procedural memory, for instance, allows us to remember how to perform physical actions. My mother-in-law was playing piano beautifully even as she suffered severe vascular dementia and couldn't remember names; her fingers remembered even as her words slipped away. That is procedural memory at work.

Semantic memory, on the other hand, lets us recall facts, names, and concepts.

In 1492, Columbus sailed the ocean blue.

I wasn't there yet I know it! That is semantic memory.

Working memory operates differently. It is more akin to RAM in a computer, storing information temporarily so we can perform immediate tasks. If I ask you to add up the numbers 5, 12, and 19, you will need to hold those numbers in your head while you perform operations on them. It's a process that requires short-term memory storage.

Episodic memory, on the other hand, is part of our long-term memory – our hard drive, if you will – and consists of the ability to recall events in their chronological, social, and emotional context. It is foundational to storytelling. You can't have a story without a recollection of events.

The question of whether animals have episodic memory is controversial. Obviously, we can't ask animals about their life stories – if we could, that would settle the issue quite nicely. What seems like feats of memory can often be explained in other ways. If a mouse eats a toxic substance, gets sick, and learns to avoid the substance in the future, that doesn't have to be episodic memory. It could be a simple association instead.

Association is a very elemental response that humans share too. For instance, if you've recently had sushi-related food poisoning, even the *smell* of sushi might repel you. It's not that you're remembering the unfortunate episode in detail, you just have a gut-level aversion. Even bacteria have been shown to be capable of this sort of learning.[44]

Some scientists believe that episodic memory developed in higher primates. In the politically complex world of a chimpanzee troop, keeping track of the behavior and performance of fellow troop-members would certainly give a chimp an edge-up on coalition-forming.[45] Others place the inception of episodic memory at earlier points in our evolution.

What is inarguable is that episodic memory is an important cognitive tool. It allows us to remember successful tactics and useful allies, danger zones and treacherous characters. It also lets us plan: by remembering our intentions we can follow through and act on them in a deliberate and strategic way. Episodic memory might also help us imagine a future by compositing memory events together.

Much of what we know today about memory was learned from one very famous experimental subject: Henry Molaison, better known by his anonymized initials, H.M.[46]

Henry

When he was 9 years old, Henry Molaison was knocked over by a boy on a bicycle, gashing his head badly. This injury may have been the inciting cause of a series of debilitating seizures that continued and worsened until, at the age of 27, they were impeding his livelihood as an electric motor repairman. In 1952, he went to the Hartford office of Dr. William Beecher Scoville, a patrician neurosurgeon related to the author of

Uncle Tom's Cabin. After exhausting all other treatment options, Scoville proposed a surgery that he would later admit was "frankly experimental."

Since the brain has no pain receptors, Scoville only needed to use scalp-deep local anesthetic during the initial incision. With Henry sedated but fully conscious, Scoville peeled the skin of the forehead down over his eyes, drilled two silver-dollar-sized holes through his skull and – with a spatula-like retractor – lifted the frontal lobe out of the way so that he could access the limbic area of the brain. Then, with a silver straw, Scoville proceeded to suction out about two-thirds of Henry's hippocampus.

Scoville's surgery actually did reduce the frequency of the seizures – but it also created history's most famous amnesiac. The memory-formation processes in Henry's brain were permanently damaged: although he could remember events from before his operation, he could form no new memories. He could cite the U.S. presidents before his operation, for instance, but he couldn't name any after his date with Scoville. Nurses who cared for him for years were greeted each morning as if they were complete strangers. He could read the same issue of TIME over and over again for weeks because he forgot both the content of the articles and the fact that he had read them.

While for the rest of us time is a vast landscape, for Henry it was just the thin 30-second slice his short-term memory could retain. Until his death Henry would need care – first from relatives and, later, in an institutional setting.

Reader, perhaps you have already gleaned that I am prone to memoir-ish asides, so I hope you will indulge me once more. You see, I was named William after the surgeon who operated on Henry's brain. Let me remember for you how that happened and then we can return, soon enough, to the theme of memory.

Quadriplegia Strikes

The story begins in Lima, Peru — it is 1960, before I was born. Jorge A. Gadea, a young father of four, is diagnosed with a herniated disk. A respected local neurosurgeon, Dr. Fernando Cabieses, operates. Within days the patient's condition deteriorates, and he is soon paralyzed from the neck down.

Cabieses didn't know it, but the problem wasn't a herniated disk. A congenital cyst, present in my father's spinal column since before birth, had begun to grow and pressure the cord. The surgery gave the cyst space to expand, compressing the nerves and causing the paralysis.

In my father's case, paralysis did not mean a lack of sensation. He later described those months of quadriplegia as excruciatingly painful. Neck traction was the only thing that relieved the painful spasms. Dr. Cabieses had no good answers, so he suggested that my father visit a renowned neurosurgeon in the United States, Dr. William Scoville.

Perhaps Scoville's entrance into this story merits an introduction.[47] The written accounts I've found of him as a younger man describe him as exceedingly handsome – and photographs confirm it. He was a high-spirited man who met his first wife by jumping onto the running board of the moving car she was in. A colleague recounts other daredevil acts: how he climbed a cable to the top of the George Washington Bridge at night, and how another time he jumped into a Spanish bullring with a live bull. As a young man, Scoville aspired to be a mechanic, but his father persuaded him to pursue the "proper" profession of medicine. His mechanical proclivity led to his inventing and refining many neurosurgical instruments, but cars remained an enduring passion; he loved working on them and he loved driving them – fast. He owned a string of Jaguars, which he kept shiny and in excellent condition.

His adventurous spirit extended to his professional life as well. The same colleague described him as "a free spirit, unfettered by rules and regulations... driven by an insatiable ego, seeking better ways of doing things and espousing new ideas with their frequent and often angry confrontations." By 1960, he was the head of neurosurgery at Hartford Hospital.

That was the year my mother and father flew to Hartford, Connecticut to see Scoville, enlisting financial assistance from my extended family, my father's Masonic lodge, and a generous American church to do so. Scoville's initial diagnosis was the same as Cabieses': that this was a case of a herniated disk. But Scoville had the audacity and skill to keep exploring until he discovered the grape-sized cyst that was causing the problem. He successfully removed it.

The day after the operation my father was taking his first steps in months. He would soon make a close-to-complete recovery, with only a little weakness in his right arm and a vertical scar on the back of his neck left as evidence of his travails. He must have been moving pretty good because within a few months, I was conceived.

Upon my birth I was named William in honor of the doctor who gave my father his life back. In our family, Scoville attained the status of a demi-God, and I imagine he enjoyed receiving that kind of esteem. He visited us in Peru during the mid-60s, and in the Dominican Republic when we were living there in the late 70s.

I was too young to remember that first visit to Lima, but many times during my childhood I heard my father tell of how he took Scoville to the Inca museum on that visit. Apparently, Scoville was fascinated by one exhibit in particular: pre-Columbian skulls with circular incisions on them that had healed. It seems the Incas – who had not yet developed writing – were performing brain surgeries... and the patients were surviving these surgeries! Scoville speculated that, based on the location of the incisions, the Incas were operating on personality. He was enthralled, and the fascination was lasting. He came to acquire and collect Incan surgical instruments.

Knowing what I've learned about Scoville, I wonder whether he was a little envious of the Incas.

Psychosurgery

You see, Scoville was one of the pioneers in bringing the lobotomy to the U.S., one of the more dismal chapters in 20th-century medicine. Before rushing to judgment, however, we need to place his efforts in proper historical context. In the 1950s, psychosurgery was considered serious science. A decade earlier, António Egas Moniz was awarded the Nobel Prize for pioneering what was then called the leucotomy. At the time, before the advent of pharmaceutical alternatives that eventually replaced lobotomy, mental patients were often crammed into overcrowded wards, enduring terrible mental anguish. Lobotomy was viewed as a humane intervention for these afflictions.

Also, Scoville was far more sensitive to the side effects of lobotomy than many of his colleagues were. While the huckster Walter Freeman was streamlining the procedure into a 10-minute-long ice pick poke through the eye-sockets, which he even performed in hotel rooms (disabling patients that included JFK's sister Rosemary), Scoville was refining what he felt was a subtler, less damaging method that he called orbital undercutting.

Yet Scoville was complicit in the big problem with lobotomy: it was the lobotomists who decided how successful they were. Even in our day of evidence-based medicine, clinical trials for new surgical techniques are a rarity. That's problematic, but less so when the results can at least be measured reliably; an orthopedic surgeon testing a new knee operation, for example, can assess post-operative flexibility and strength down to precise degrees and pounds of pressure. Measuring psychic pain is far more difficult. And crucially, weighing alleviation of pain against loss of intellect and higher functioning is exceedingly difficult. That is as much a philosophical as a medical problem, and that is what psychosurgeons needed to decide. Such a subjective call should not be left to doctors with a possible attachment to their handiwork.

There is evidence to question whether Scoville's judgment was always sound in this regard. On Henry Molaison's chart prior to discharge, Scoville marked his patient's condition after the operation as "Improved." And in a 1960 paper he writes that total lobotomy is more appropriate for patients with a "low cultural background."[48] Presumably, this is because he believed that people with a lower educational level were less likely to miss the higher intellectual capacities that a lobotomy might damage, but really… isn't that an outrageous presumption?

Scoville didn't just start performing psychosurgeries earlier than most, he held onto their practice longer too. By 1973, lobotomies had almost disappeared, rendered obsolete by drug therapies and electro-shock treatment, but a *New York Times* article of the time introduces Scoville as having "possibly performed more lobotomies than any other surgeon alive today."[49] (Freeman had died the year before.) Even at that late date Scoville remained a proponent, claiming a 50% improvement rate, while admitting there is a price to pay: "the blunting of higher sensibilities, such as intelligence, abstract thinking, and the ability to fantasize."

Of course, my family never knew or talked about Scoville's

lobotomy operations, or the case of H.M. I didn't find out about it until the mid-90s. I was leafing through my Sunday paper when Scoville's name ambushed me from out of the page. It was a review of Philip Hilts' *Memory's Ghost*, a book about H.M. The review made Scoville seem craven, even joking to his wife about his mishap. I remember so clearly thinking: 'I wish Papi were alive so I could see what he thought of this!' It's only when I was writing this that I constructed a timeline and soon realized... that thought would have been impossible. The book came out in 1995; my father did not die until three years later.

Which leads us back to memory: how does it get bent, as I bent it? Suzanne Corkin, the scientist who conducted much of the research on Henry during the latter part of his life, writes about how the process of reconsolidation helps maintain memories. But she also says:

> *Think of reconsolidation as a memory-updating process. If you unpack a suitcase and then repack it, the clothes will be arranged slightly differently than they were before, and you may leave out some items and add others. Retrieval and reconsolidation of old memories suddenly makes them labile, a state in which they are again vulnerable to distortion and interference.*[50]

That's helpful, but it doesn't answer the question of why I didn't put *this particular* garment back into the suitcase. I probably didn't tell my father about the H.M. operation because such an act might be considered an attempt to topple Scoville's statue off its pedestal. Such an affront to Scoville might be upsetting to my father, or awkward for me. So, I chose not to tell him.

And years after I made the decision not to tell him, I chose to think that he had died by that time because the decision not to tell him made me feel small – I would have to admit to myself that I was fearful of my father's reaction. I prefer to see myself as strong and independent, so I changed the memory to conform with that self-appraisal. I changed the story so it would be more flattering to me.

Henry was unable to craft a personal narrative in this way. The story of our Atomist Self is a tale of failures and triumphs – trials passed and lessons learned. It is a story that trails behind us but also extends beyond us, piercing the future with the glory or oblivion we imagine for ourselves. Henry had none of that.

Scoville certainly did. I'm sure his dreams were of beneficent glory, but I suspect they sometimes overshadowed the frail specimens of humanity before him. The personal narrative that was forever quieted in Henry must have been a raging, heroic symphony in William Scoville's head. It must have made the wispy music of the world sometimes hard to hear.

In The Moment

So what was it like for Henry to live in the eternal present? Corkin reports:

No one would doubt that Henry's experience was a tragedy, but he rarely seemed to suffer and was not continuously lost and frightened – quite the contrary. He always lived in the moment, fully accepting the events of daily life. From the time of his operation, every new person he met was forever a stranger, yet he approached each one with openness and trust. He remained as good-natured and pleasant as the polite, quiet person his high-school classmates knew. Henry answered our queries patiently, rarely getting angry or asking why he was being questioned.[51]

Perhaps living the way Henry did is something we should aspire to? Corkin, who spent years with Henry, is willing to entertain the possibility:

As frightening as it seems to live without long-term memory, a part of us all can understand how liberating it might be to always experience life as it is right now, in the simplicity of a world bounded by thirty seconds... Dedicated meditators spend years practicing being attentive to the present – something Henry could not help but do.

There was a phrase that was so characteristic of Henry that Corkin uses it as a chapter heading in her book about H.M., *Permanent Present Tense*. When Henry was uncertain, he would sometimes say: "I'm having an argument with myself." It's such an unusual phrase. I wonder if the weakening of his personal narrative might have loosened his sense of a unitary Self, pulling back the curtains on the debates in his Parliament of Mind.

I hesitate to poke at the corpse of Henry Molaison, hoping to extract one more insight from this grand case study, but thankfully, this will only be – and can only be – a thought experiment.

Imagine yourself in a modified version of the state Henry found himself in – you will lose some aspects of your memory as he did, but not all of them. Unlike Henry, you will be able to remember and track your personal relations, so that new people aren't always strangers.

Also unlike Henry, you will retain your memory for facts. Henry was able to form new procedural memories (i.e., learn new physical skills,) as Brenda Milner demonstrated in the mirror-drawing experiment she conducted with him.* This proved that the faculties of procedural memory (which he retained) and implicit memory (which he did not) resided in different parts of the brain.

You, however, will retain both abilities. You will be able to learn new physical skills *and* you will be able to remember episodes of your life or new facts perfectly.

What you will not be able to do is remember a personal narrative of yourself. You will not be able to remember your strengths and weaknesses, or what you consider to be your essential traits. You will not be able to remember how you see yourself in comparison to others. And while you will remember events, you won't be able to remember an interpretation of how those events shaped you. And crucially, you won't be able to remember an imagined future. You will not be able to build a palace of dreams in your head.

If some psychosurgeon of the future could safely and precisely create this mental state, could we attain some sort of instant enlightenment this way? Or, to pose the question in a less loaded way: is a life without a personal narrative one to recommend? Is it a life *you* would choose to live? Pick your eyes off this page and consider this question for a moment, if you will.

---

* In this experiment, Henry was asked to trace a star shape, which he only saw in a mirror, reversing the visual feedback. This is a lot trickier than it sounds, but once you have learned the skill it stays with you. When Henry was asked to repeat the task at a later date, he had no memory of having done it before – yet his hands remembered. The skill he had learned was still there, even if the memory of learning it was not.

Done? Well, I don't know about you, but my answer comes quickly and easily: no, it is not.

I cannot speak for the Buddha, of course, but based on the story below of how he came to his awakening, my suspicion is that he would agree.

The Ascetics

After the Buddha – then known as Siddhartha – abandoned his everyday life, he sought teachers to show him the way to liberation. He found a teacher from a prominent spiritual movement of the time known as the *śramaṇas*; we commonly refer to them today as the Ascetics. These teachers demanded a complete abandonment of worldly life: cutting off familial ties, material possessions, and all the trappings of status.

For Siddhartha (the Buddha's name before enlightenment), these weren't trivial sacrifices. He was a wealthy prince, married to Princess Yashodhara and expected to inherit the throne of King Suddhodana. On the very day he left his home to join the Ascetics, his first-born son, Rahula, was born. Does the act of abandoning your wife and new-born seem selfish, self-absorbed, and unworthy of a spiritual savior? Perhaps it is. Or perhaps in Siddhartha's mind the redemptive power of the future enlightened being he imagined himself becoming outweighed the current harms to his family.

The Ascetics taught Siddhartha not only about meditation, but also that he must overcome attachment to his body and its physical sensations. To do this, they practiced strict fasting. This was meant to weaken the body and its demands, and thus purify and strengthen the spirit. It is said that Siddhartha lost so much weight that a person could wrap their hands around his waist and touch the thumb and forefingers of each hand to the other. He stumbled around this way, weakened but in desperate search for truth.

Something about this path did not seem right to Siddhartha. As he sat frail and weakened, a peasant girl named Sujata passed by and saw his diminished state. The sight of him stirred her compassion, and she offered him some rice pudding, which Siddhartha gladly accepted. The food strengthened his body and will. He came to a firm resolution.

He would sit under a tree and not get up until he was enlightened. By the time dawn came the next morning, he found what he was looking for. It was not the absolutist self-abnegation of the Ascetics, and it was not the blithe sleepwalking of his life before. It was a Middle Way – a method of seeking liberation while participating in the sorrows of the world.

Years later, he would be reunited with his son Rahula, who would become one of his foremost disciples.

To abjure from self-concept, to give up plans and ambitions, to not plumb our personal history for lessons and learnings – this feels to me like the world-denying path of the Ascetics, rather than the life-embracing path of the Gautama Buddha.

However, I should admit that my answer to the thought experiment I described above was not always so confident. There was a time when I thought that my practice did require self-denial. I thought that erasing my personal narrative and subduing my ambitions was what my Zen practice asked of me. I went to interview with my teacher, Enkyo O'Hara, and asked how I could quell these dreams of achievement. She looked momentarily daunted by how confused I was, and then she smiled and said simply: "That's just Boddhisatva energy."

In Buddhism, a Boddhisatva is someone on the road to Buddhahood. However, instead of storming on to Nirvana, they hang back to help others. Roshi Enkyo – a woman who founded a vigorous sangha* in the middle of Manhattan, and whose many dharma successors have spread the teachings widely – is a woman of ample ambition. She would never have discouraged it in others.

Our task is not to lose our stories or our aspirations. Our task is to continuously remember that they are just stories, just mind matter.

Epitaphs

Mind matter or not – it's best to finish the stories you start, so I shall recount the rest of my experience with Scoville.

In 1979, I was 17, living in the Dominican Republic and very excited about going off to college at NYU. I was literally counting down the days in my head. Around the time I was down to single digits, the Dominican Republic was hit by one of the worst tropical hurricanes of the century: Hurricane David. Our ground floor apartment was flooded with a foot of water. Looking out the taped windows it seemed like I was peering into the contents of a blender. Trees were uprooted within our view. The next day we would learn that the eye of a Category 5 hurricane had passed almost directly over us. 2,000 Dominicans had been killed.

My father was working for a non-profit at the time. It was quickly decided that, to solicit funds for hurricane relief at various agencies, he would accompany me on my move to college. He had recently been feeling some neurological symptoms: he could walk but he felt unsteady and appreciated an arm to lean on when he negotiated a curb, for instance. We would take advantage of my trip to New York to get my Dad a

---

* A sangha is a Buddhist community of practice.

check-up with Scoville at Hartford Hospital.

Dr. Scoville greeted us in a warm but brisk manner. After running some tests, his verdict was curt: "I'd like to operate again." He explained that he would need exploratory surgery to see what was going wrong. My father wasn't expecting this. "Do I have time to return to Santo Domingo and put my affairs in order?"

"I'd prefer to do it sooner rather than later," Scoville replied.

The operation went for longer than expected. Being in the waiting room felt a little surreal... like I was in a too-melodramatic movie. I remember a kind nurse asking if I needed anything. And my memory of this is hazy, but I think that after many hours Scoville came out and said something non-committal.

Scoville had only found adhesions, the scarring from previous operations, which he tried to remove as well as he could. However, my father woke up largely paralyzed from the neck down and would remain so for the rest of his life. He would retain limited use of his hands and a bit of feeling in his legs.

Years later, after Scoville's death, my father would travel to Los Angeles to be examined again – this time with the benefit of an MRI, a diagnostic tool not commercially available at the time of Scoville's operation. The doctors would tell him that Scoville had operated in the wrong place. The cyst had regenerated, but in a new location. Scoville had missed it altogether. Perhaps the operation itself did damage to the spinal cord, or perhaps it aided the progression of the cyst, as Cabieses' operation apparently did. Either way, it seems likely that a better outcome was possible.

Looking back on those days since, I've often thought: if the boy that accompanied my father to Hartford Hospital had been a man, he might have had the presence and maturity to worry whether an operation that so depended on motor skills should be left to a septuagenarian. He might have had the sense to suggest that they seek out a second opinion, even if Scoville was good. Truthfully, I don't feel guilty about the decision I did not contribute to — the decision to operate in Hartford with Scoville. I find it highly unlikely that the 17-year-old boy could have stood up to the family God.

How should we appraise the doctor from whom I got my name?

William Scoville coveted the power to reshape the brain, the seat of the soul, the source of our suffering. He sought the power of a God. And like Icarus and Prometheus and all the other fellows who got too big for their britches, in the case of Henry Molaison he had a comeuppance — a very gentle one by mythological standards: no plunge into the sea, no eagles gnawing at his liver. The price he paid was that – despite having a long and distinguished career – this single botched operation was what he mostly became famous for. (It was a gentle comeuppance by psychosurgical standards too: the Nobel-Prize-winner Moniz was shot and left paralyzed by a patient he operated on; Freedman lost his medical license after a woman died during a procedure.)

I cannot judge him harshly. The boldness and over-confidence that caused Scoville to cripple Henry is the same boldness and over-confidence that allowed him to cure my father... and allowed me to enter this world. Without the arrogance of William Beecher Scoville, without the world-filling deluge of his personal narrative, you would not have this book in your hands. (And of course, to write a book on these subjects that I am not at all credentialed in requires a good bit of my own arrogance, too.)

Scoville was killed in an automobile accident in 1984. He was backing up on a highway to catch an exit he had missed and was rear-ended by another car. Bold to the last.[52]

My father spent the last 19 years of his life in a wheelchair. He died in 1998 from the same thing that killed Louis Leborgne and many other paralysis patients: complications from bed sores.

Henry Molaison — the famed H.M. — outlived them both. He died in 2008, sparking many wonder-filled obituaries.

CHAPTER 6

THE WORLD WITHIN

Let's return – once more – to our ancestor's fireside story that began this book.

When Shiki rises to tell his tale, his listeners – gathered around the fire – enter into deep syntony with him. They picture the deer in the field. They picture the ravine where two hunters lie hidden. They picture how they will fan out in a semicircle, driving the deer toward the narrow pass, culminating in the bounty of a successful hunt.

How do they do this? How do *we* do this? Memory provides the raw materials. But then our mind gets to work, manipulating these fragments to create scenarios we've never experienced in quite that way before.

When – and how – did we develop this astonishing ability to create mental simulations? And do any other animals share it?

As noted in Chapter 2, one basic function of consciousness is to synthesize sensory input into a coherent model of the world. This allows us to derive more complete and useful information about our environment. Pooling sensorial information enhances its value. But mental simulation goes a step further: it lets us rearrange fragments of memory into something new – a mental model we can test, explore, and learn from.

For example, to use tools in a novel way you've never seen before, you first must imagine how those tools might solve a problem in physical space. You need to run a simulation and experiment with it in your mind.

Chimpanzees have often been observed using sticks to fish for termites or using rocks to crack nuts. And they're not alone: dolphins, otters, and birds – among others – have also been seen using objects as tools in the wild.[53]

The tool skills of animals are typically learned by imitating their peers, but in a captive experimental context, apes have discovered truly original solutions to problems. In one experiment, for instance, an orangutan faced with an unreachable peanut, trapped in an upright plastic tube, used his mouth to transport water to the tube, filling it until the peanut floated to his reach.[54]

Tool use reveals something else essential to storytelling: causal inference. To use a tool, one must grasp that *this* action will lead to *that* result. And since a story is, as we have defined it, a representation of a causal sequence of events, there can be no storytelling – or story-receiving – without this capacity.

The Predictive Model

However, if we consider that manipulating a mental model of the world is a fancy skill that only a handful of species have, then we might miss something fundamental about how our mind works. A well-supported theory suggests that our consciousness is not a live feed, so to speak. It might seem like the data of our senses is constantly flowing in, but our senses provide a firehose of information that would quickly overwhelm our ability to process it efficiently. Instead, many neuroscientists believe we have developed a predictive process. We use previous experiences to create a model in our minds of the world we expect, and then as new sensory information comes in, we test this model. If it is in error, we make necessary updates.[55]

One point of evidence to support this theory is how the blind spot in our vision is filled with a view, instead of black. Here's how to identify your blind spot, if you haven't tried it before. Close your right eye and focus on the circle in the following graphic. Bring the page close to your eye – almost a nose-length. Now slowly pull the page back. At some point, the cross will disappear.

That is because the cross is being projected into a part of your inner eyeball where the optical nerve enters, and there are no light receptors. You can't really see what is in that part of your vision. Yet, when you take the page away, still with one eye shut, you don't see a black spot there at all. Your brain has filled in that area with its expectations.

There are more baroque examples of how the predictive model works.

Many of us have had the experience of thinking we recognize a person when we are out in a crowd. We come a little nearer, and once we have more visual information, we see that it is not our friend but a total stranger. At that stage we update our assumptions and realize that there wasn't that much of a similarity to begin with. That our acquaintance would never wear those pants, for God's sake, and that we were simply filling in details that weren't there.

Scientists have recreated this experience in experimental settings with visual illusion experiments; they have scanned the brain while introducing break-in-expectation stimuli. The evidence they have gathered hints that our consciousness acts as a sort of artist in our head. This artist keeps an eye on the scene she is painting, and if it matches what is in her painting she holds back her brush. If, however, a bird should fly into the frame, she quickly paints it into the picture. This whole process takes just a few milliseconds and continues for as long as we are conscious.

Hildy's Ball

Of course, it is not necessary to credit the predictive model theory to believe we create a simulation of the world in our minds. What else could it be? I am looking at my cat Hildy's toy, a small red ball, right now. The image in my mind is obviously not the thing itself; it is a representation of the ball.

The process of seeing Hildy's ball goes something like this: light is reflected off the ball, it is focused by the lens in my eye and projected onto the retinal layer at the back of my eyeball, where light sensors in the shape of rods and cones pick up the light and initiate a signal via the optic nerve to the visual cortex in the brain. The brain recognizes a circular outline and the tell-tale shading pattern of a sphere: radially gradated light from bright to less bright on one side, and shadow on the other. Classification is an integral part of consciousness that allows us to set up our mental model. Hildy's ball at this stage is recognized as a sphere, and a circular object is imbued with volume and shape. (The stereoscopic action of our two eyes also helps in this task.) Because of the context, this particular sphere is recognized as a cat toy. In another context, the sphere might be recognized as a baseball or an orange or a light fixture.

Our classification of sensory experience is not confined to shape, of course. Our experience of color is a form of classification, too. I see the ball as red because its material has the property of reflecting light with a wavelength of approximately 630 nanometers. The attachment of the experience of red with that wavelength is somewhat arbitrary. The experience of color is just an identifier for the wavelength of the light, which in turn is an identifier for the sort of material that reflects it – literally a form of color coding.

It's interesting to speculate whether the experience of color is the same for all of us. If my color wheel were spun around and I experienced red as green and green as blue and blue as red, would this be discoverable by me? By anyone? Every time you see something red I would see green and call it "red," because that's what I was taught to call that color. It's true that there are qualities we ascribe to colors – such as warmth or coolness – that seem to be universal, but perhaps those come from connotations. Maybe the green I see instead of red will seem like a warm color to me because I associate it with green-hot coals! And I will think of red as a cool and soothing color because I associate it with the clear and cloudless red sky.

The idea that what our senses experience is not real is an integral part of Buddhist thought. Many texts, most famously the Heart Sutra and the Diamond Sutra, talk about the illusory nature of the senses. There are many variations of the idea that life is illusion, but the most robust version of this philosophy teaches not that the world does not exist, but rather, that our experience of it is illusory, i.e., not the thing itself.

There certainly is a small mass of plastic on the floor in front of me, which I call Hildy's ball. Its presence is communicated by reflected light into my eyes and this signal is taken to my brain, where it is assigned properties: sphere, red, cat toy, belonging to Hildy. If I should tell a story that includes this ball – how I threw it one day, it bounced off the wall, and Hildy jumped onto the wall and sprung from it like a parkour artist to follow its trajectory – then what I am telling you is a symbol of symbols: a whole concatenation of symbols really, that you are able to absorb, and you will miraculously recreate the experience of my consciousness in yours – you will receive the story.

However, there are limitations to this experience-transfer. Thomas Nagel in his essay *What Is It Like to Be a Bat?* concluded that it was impossible to know another's subjective experience. Bats use echo-location to know where they are in space. There is simply no way for us to experience how that information is coded for a bat. Do bats feel near objects (as opposed to far ones) as if they were red, or loud, or warm, or perhaps as some other novel sensation that we could never dream of? Similarly, I can't experience your experience, and you can't know whether what we both call blue I actually experience as what you call red.

If we were to shear subjective experience from the world, what would be left? There are some plastic molecules that make up Hildy's ball, but they exist in a cold and symbol-less place. Of course, that is not to say that this experience-less world doesn't have consequences to our organism. It contains buses that might mow us down if we do not heed them, bodies of water that might drown us, nourishment that we need to remain alive. But it is silent and dark and neutral; it is not the world we live in. What we know, and the only thing we'll ever know, is this beautifully designed user interface to the world that we call our consciousness.

Road Trip

Let's set aside this idea of consciousness being a symbolic representation of the world – we will return to it soon. For now, let's imagine that sitting on a cushion to meditate is like taking a road trip in a car (i.e., your body).

Give the usual person you call yourself – which we have been calling your Atomist Self – the passenger seat for a minute. All those stories, all those self-definitions, all those plans… to the passenger side, please. With that Self to your side, what is left? Who is in the driver's seat? Well, you are, of course. Who is you?

The experience of holding this book in your hand – be it in electronic or print form – is you.

The sensation of your seat beneath you is you.

The air you breath in – and out – is you.

The gurgle of your last meal in your stomach is you.

The echo of these words in your head is you.

Now look to your Atomist Self beside you. Could you let your Atomist Self do what it does best – which is mind-wander – but observe your mind while you are doing it? Try it now. I don't mean just letting your mind chatter, remembering what you were thinking, then analyzing it, but rather being a present observer of your wandering mind – functioning on two parallel tracks: an observer and a mind-wanderer at once.

I admit, I cannot watch my own mind-wandering. It's like I take a step and my foot gets stuck in deep mud. Perhaps the act of observing yourself is goal-oriented enough that the see-saw in our minds tips away from the DMN side, and it's hard to get it to tip back the other way unless you remove the weight of observing.

At any rate, this is what happens in meditation. You're looking ahead at the road, paying no heed to your DMN, and the Atomist Self starts to babble: *Hey, this meditation is going pretty well, huh? Or it was until I piped up… haha. What are we going to eat after this? Are you hungry? Hope it's not snowing out, or getting home is going to be a mess.* You cast an eye sideways at your Atomist Self, and all of a sudden he/she turns shy and quiet. But turn your eye back to the road and soon the babbling starts again.

If you have ever tried meditating before, you recognize this pattern. On ordinary days it's pretty much all we experience meditating. But let's say you attend an intense period of practice known as a *sesshin*, or retreat. You might have a whole day, or two or three, of the experience described above. But at some point, perhaps, something distinct starts happening.

You realize that this whole metaphor we are engaging in is preposterous. You are noticing yourself: you are subject. You are noticing yourself: you are object. You are both subject and object. There is no Atomist Self – that is just a concept you plucked out of some book. Is there a real you? Well, there's experience, isn't there? The breath you are noticing going in and out, the mat's pressure on your legs, there's certainly those things… but there is no separate mind-wanderer and observer. Those are just categories you applied to your conscious experience, and after some time on the cushion those categories have fallen away. Are there some distractions still? Sure, there's some, but they are part of the same landscape. They don't have a voice and an identity of their own – as soon as you notice them they are not distractions.

So there's no driver and passenger any more. How about

the car (i.e., the body)? Well, the breathing in and out, the pressure on your legs – these are all things that are happening in your body. These bodily sensations are what we are observing while we are on the cushion. The car is the object. But who is the subject? Surely it's also the car; your mind is in no other place than your body. Again, subject is object, object is subject.

At this point, we go back to what we talked about at the beginning of the chapter. All these experiences – your thinking, your emotions, your bodily sensations – all are happening in your consciousness. But how about the world outside your body? That is in your consciousness too. Everything inside and outside of you is part of the same canvas. They are just different neurons firing. Of course, it is always this way, but the practice of sitting denudes our experience from the veils of thinking that we drape over them; practice allows reality to come forth and exhibit itself plainly.

When this happens, not only does the division with your passenger and the car disappear, but the division with the countryside outside the car disappears as well. Everything that is happening – driver, Atomist Self, car, countryside – is happening inside you. You don't need to harbor the illusion of an inner observer to see it.

Ironically, we've used witness mode to get out of witness mode.

You can take this as a description of what is going on inside a meditator's psyche, but it is also the simple, undeniable, physical truth of the matter: subject is object, object is subject.

The Old Joke

Have you heard that old joke?

What did the Buddhist say to the hot-dog vendor? Make me one with everything.

Not to be a pedant, but that's not quite right. It's not really like being one with everything, it's more like letting go of counting... even to one. Some people also describe the experience this way:

Not one, not two.

There is still a car and a countryside; it's not like they blend into one blob. Your perception of the world is no different than it is in your everyday life. You are not armed with a new psychedelic vision. But your relationship with the world, how you regard it, changes distinctly. You are no longer dividing the world with mind.

You've probably picked up by now that there are two ways to consider the dharma.* One way is to understand it intellectually. If you've followed the argument above, you can say you understand a very important aspect of the dharma. Congrats! Release the streamers and balloons.

How much does that realization count for? In Zen practice, we are not looking for intellectual understanding. We are trying to live this truth – to experience it. The experiential exploration of this reality is what changes lives; intellectual understanding counts for much, much less. But the good news is, experiencing it is easier than people think. Of course it takes sincere application. But this is not something that is only the province of gifted mystics living as hermits on top of a

---

* This is the Sanskrit word that Buddhist use to refer to the Buddha's truth.

mountain for years. You don't need the luck of a lottery-winner to get there.

Robert Wright, in his fine book *Why Buddhism is True*, writes about an experience he had at a Vipassana retreat:

At one point I felt a tingling in my foot. At roughly the same time, I heard a bird singing outside. And here's the odd thing: I felt that the tingling in my foot was no more a part of me than the singing of the bird.[56]

He later writes:

I've had several chances to describe the experience to truly accomplished meditators—some of them monks, some of them famous meditation teachers—and invariably they've recognized the kind of experience I'm describing as one they've had.

I don't count myself as a particularly accomplished practitioner, but I'll raise my hand here! Yes, that is an experience I recognize. And plenty of reasonably dedicated lay practitioners report having experienced something similar. It's not a big deal. If you haven't experienced it, you shouldn't think of it as a goal, and if you have experienced it, you shouldn't think of it as a trophy. There's a phrase in Zen that has been interpreted in many ways:

If you see the Buddha on the road, kill him.[57]

The way I would interpret this is: don't cling to these moments. They are just something else that happens to you.

Next week you might well be cursing silently at your boss again. Your practice continues.

While it would be counter-productive to glorify and romanticize these experiences, to portray them as undifferentiated from each other would be arrogant and lacking in humility. It would be an error in the opposite direction.

When I go to my cushion for my daily sit, perhaps the door to the Connected Self opens a tiny crack. When I go on retreat, perhaps that same door comes ajar and a sliver of light shines through. For many gifted and/or dedicated mystics (and I am definitely not in this number) the experience is far grander. History provides ample testimony of practitioners from many different traditions, for whom the door to the Connected Self swings open violently, letting in blinding light and gusts of overwhelming emotion. These are the stories of Saul on the road to Damascus, Joan of Arc's visions, and the Buddha's awakening. But tales such as these are not confined to history's boldface names. William James collected boatloads of such mystical testimonies in his invaluable *Varieties of Religious Experience*.

Measuring Transcendence

Andrew Newberg has dedicated much of his professional career to studying how religion and neuroscience interact. To study religious experiences, he sought to measure the associated brain activity, just like Raichle did when he discovered the Default Mode Network.

However, Newberg realized that being in an MRI machine – flat on your back in a claustrophobic tube, surrounded by high-decibel whizzing – wasn't ideal for prayer or meditation. So, when he sought to study contemplative practices, he chose a method we haven't yet described: the

SPECT camera.

In this technique, the experimental subject is injected with a tracer liquid containing a low level of radioactivity. This tracer bonds quickly with active brain cells and remains there for hours. Later, without any rush, the experimenter's camera can capture that moment in the past when the tracer first entered the brain, identifying where the substantial blood flows occurred at that time.

Newberg describes his experimental process at the beginning of his book, *Why God Won't Go Away*. A subject who is a practicing Buddhist sits on a cushion, going deeper and deeper into meditation. At the scientists' request, he tugs a piece of twine just as he is entering the deepest phase of his meditative concentration. Hidden in another room, Newberg and his colleagues feel the tug and press a button to start the marker flowing through an IV tube. That meditator's moment of contemplation is captured the moment the tracer liquid enters the brain, and it is now available to be imaged at the scientists' leisure.

Starting in the late 1990s, Newberg, in collaboration with his colleague Eugene D'Aquili, used tools such as these to observe the minds of Tibetan Buddhist monks and Franciscans nuns. While the patterns of mental activity were not identical, the similarities they observed between the two kinds of practice were striking, with similar patterns of activity in four separate brain areas.[58] Interestingly, they noticed a drop-off in activity in an area responsible for orienting the body in space, what he called the Object Orientation Area. He speculated that perhaps this deactivation caused "the sensation of the self [as] endless and intimately interwoven with everyone and everything."[59]

Newberg calls this mystical state of mind Absolute Unitary

Being. There is no difference between that and what I've called the Connected Self. Newberg proposes that this experience is cross-cultural and universal:

> Neurologically, and philosophically, there cannot be two versions of this absolute unitary state. It may look different, in retrospect, according to cultural beliefs and personal interpretations – a Catholic nun, for whom God is the ultimate reality, might interpret any mystical experience as a melting into Christ, while a Buddhist, who does not believe in a personalized God, might interpret mystical union as a melting into nothingness. What's important to understand, is that these differing interpretations are unavoidably distorted by after-the-fact subjectivity.[60]

Maybe the same sort of things are happening in the brains of Catholics and Buddhists, but since our beliefs are so different, surely our practices are quite distinct, right? Let's try a thought experiment. If we invited Martians to visit my zendo during practice, and then visit a Roman Catholic church during mass, how would they consider these two events, based on face value alone?

The Martians might notice that as the Zen practitioners enter the zendo for a session, they give a bow and, in some Zendos, also enter with the same foot. As the Roman Catholics are entering church for service, they will bow and cross themselves. *Religions usually mark a sacred space and time for practice.*

Both the Buddhist and the Catholic assemblies would have an altar with a sculpted figure at its center. *While the depiction of divine forms is sometimes proscribed, religions use visual art to heighten the spiritual experience.*

At some point in both the Zen and the Catholic ceremonies, people would prostrate themselves. In both, the congregants would bring their palms together with fingers pointed upwards, in what the Catholics call *orans* and the Buddhists call *gashho.* In both ceremonies, there would be singing or chanting. In both, movements would follow pre-arranged patterns. In both, there would be candles and incense. *Religions use rituals containing coordinated movements, defined hand positions, prostration, speaking or signing in rhythmic unison, and sometimes aromatic elements. These actions metaphorically humble the individual and foster the experience of unity with the religious community.*

At some point, the Zen Buddhists focus their minds by observing their bodily sensations in Zazen. The Roman Catholics might inwardly recite Hail Marys while they finger a rosary. *Religions use different meditative and contemplative techniques to quieten the rational mind.*

In both the Zen Buddhist and the Catholic services there would be a talk by a robed officiant. The officiant will reference or recite sacred texts. *Religions emphasize the passing on of the word by written or spoken means.*

If our Martian friends knew nothing more than what they could observe, wouldn't they assume that the Catholics and the Buddhists were engaged in a very similar activity – even when the belief systems could hardly be more different? I believe if they came to such a conclusion, they would in fact be correct.

If our esteemed Martian friends had a scholarly bent and studied religions more deeply, what common elements might they recognize? They would probably recognize that despite diverse cultural origins, there are surprising similarities between religious cultures, and not all these semblances can be attributed to common cultural roots or influence on each other. There are four important common elements:

Religions offer a cosmology (although as we have mentioned, Buddhism's cosmology leaves out the origin story).

They offer a moral code: a way of harmonizing with the world, or, in the parlance of the deist religions, obeying the strictures of a loving God.

They offer community: a fellowship of practitioners that can act as balm, support, and good example – as well as provide a workshop for good deeds.

Religion's final and essential function, which underlies all other functions, is that it evokes the sacred: the experience of unity and self-transcendence that we have been calling the Connected Self.

Given that religion has been a part of every human culture we have discovered so far, should we assume it is hard-wired into us? It's possible, of course, that it isn't. It could be what Daniel Dennett calls a *good trick*,[61] meaning a cultural solution arrived at independently many times over because it is a sound and apparent solution to a problem. However, twin studies show a degree of heritability for religiosity that usually ranges from 30 to 50%, so it seems unlikely that genes play no role at all.*

---

* This type of study typically compares the traits of identical twins, who share a 100% of their genome, with fraternal twins, who only share 50%, to calculate how much of the trait is due to genetics. Both kinds of twins

If religion is hard-wired into us, then how could that come about? How could self-surrender lead to a survival advantage for an individual organism? How could selflessness provide evolutionary benefits?

Birds Of A Feather

Darwin himself proposed a mechanism that might explain why we would invest in others at the expense of ourselves. He called it group selection.[62] His theory was that if individuals within a group possess a trait that is self-sacrificing but benefits the group, then the group's overall fitness might be improved, allowing that trait to spread – even if it does not directly benefit the individual. The classic example of this mechanism is a flock of birds, where an individual bird utters a cry of warning upon spotting a predator. This cry might increase the individual's vulnerability slightly but, by doing so, it increases the chances that others in the flock will survive. A flock with this trait would be more fit for survival than one without it, and the self-sacrificing behavior of warning would be passed on.

This view of group selection held sway for many years and seemed to have much explanatory power. But then, in 1966, George C. Williams wrote a book called *Adaptation and Natural Selection*, which picked apart the case for it. He argued that cases such as the bird warning the others in its flock could be explained away by natural selection alone, without requiring group selection as an explanation, and he cast doubt on the mathematical models that supported the idea that traits could be passed on this way.

Group selection came into a couple of decades of

---

are brought up in the same environment, so that variable is seen as minimized.

disrepute, but around the turn of the century a few theorists rehabilitated the concept. David Sloan Wilson argued that Williams was committing what he called the *averaging fallacy*: assuming behavior in service of the group but attributing the benefit to the individuals in the group. Wilson proposed that natural selection occurs at many levels at once. Certainly, there is the genome level, which Dawkins championed, but there is also the kinship level, the group level, and the species level.

Wilson would later write a book called *Darwin's Cathedral,* which addresses the issue of religion. He proposed that the spiritual impulse is indeed an example of group selection: that religion is a co-evolution between a genetic trait and a group's culture, and gives the group a survival advantage.*

So, what is co-evolution? It is when genetics and culture evolve in parallel, each affecting the other. The most cited example of this is how human populations that breed cattle tend to develop lactose tolerance into adulthood. Wilson argues that religion, with its moral code and group identification, provides a cohesion that gives the group an advantage over those outside the group. Those with a religious proclivity join the group, and the group lends its strength to the joiners, so that this trait is passed on genetically and culturally. Crucially, the in-group out-survives the out-group.

While kinship and species levels of selection are hard to deny, group selection is still a fiercely debated proposition within the scientific community. I will not take a side in that argument, but I will say that it is not necessary to believe in group selection to believe that the spiritual urge is part of our biology, and part of our genome.

---

* A better-known scientist of the same surname, E.O. Wilson, also came to a similar conclusion.

An Evolutionary Basis

Before I explain how evolution could have provided us with a Connected Self even without group selection, let me define what I mean by Connected Self, because I realize I have been using the term somewhat promiscuously. It is what Newberg calls Absolute Unitary Being, that experience which a Buddhist might see as emptiness, or a deist might see as union with God. But it is also the behaviors that foment and abut that experience; it is the entire practice of religion. More than that, I also mean the secondary effects of practice on practitioners; what it indirectly leaves in us.

I have previously proposed that the Atomist Self is a trade-off adaptation that introduces real vulnerabilities to our organism.

We have also proposed that the Connected Self is a cultural-genetic adaptation that addresses this vulnerability.

If both statements are true, then a predisposition to a Connected Self could be acquired at the individual level. In other words, the Connected Self might make the group stronger in comparison to other groups, but it also makes the individual stronger in comparison to other individuals. Compared to a non-practitioner, a practitioner will be less likely to be exposed to vulnerabilities such as anxiety, depression, lack of confidence, contentiousness, and isolation. This makes her fitter to survive.

If it were true that religion offered an individual fitness advantage, then you would expect to see better health outcomes among religious people. Indeed, this area has been thoroughly studied, and the confirmation of this proposition is quite robust. A study of 74,000 women who attended religious services found a 33% lower risk of mortality compared to those who did not, even after adjusting for possible confounding factors[63] – that's comparable to the health advantage of quitting smoking. Many studies and meta-analyses have associated religious practice with less depression, lower anxiety, greater life satisfaction, lower cortisol, reduced blood pressure, and lower substance abuse, among other benefits.[64]

Contrary to atheist critics who believe it is just superstition that comforts us about our mortality, or cynics who think it is just an ingenious method of social control, religion is a tool to address the pains and contradictions of the Atomist Self, while also weaving together tighter-knit communities.

The Connected Self, then, is a salve to a real biological vulnerability.

When a Christian says grace before dinner, when a Muslim looks for the direction of Mecca to pray, when a Jewish woman lights the candle before *shabbat*, when a Hindu performs a *puja* for Ganesh – and when I sit on my cushion – we are all doing the same thing. We are all – even if just partly and temporarily – releasing our Atomist Self as we bolster our Connected Self. This is a method we have been taught and have learned, but it is also something we are biologically built for.

Your Original Face

In this chapter, we've described the magic of how we create a world within our mind, and we've followed that trail to the Buddhist insight that there is no separation between self and world – that it is all the same garment.

We've also seen how the thirst to experience this truth, this numinous feeling of unity, is cross-cultural and common to many traditions.

Our face has eyes, a nose, a mouth with a tongue. These features evolved because they conferred an advantage: eyes to see, a nose to smell, a tongue to taste.

There is a famous Zen koan from the *Mumonkan*, Case 23:

What was your original face before your parents were born?

If you search for your original face, you will find evolutionary gifts there too. It is the sensing organ for your truthful place in the world.

CHAPTER 7

GODS, FOUND AND LOST

Witness the birth of an important component of storytelling in a common family scene: a parent enters the kitchen and finds a cookie jar opened, crumbs on the floor, and chocolate smeared all over the mouth of her toddler. That cookie was *verbotten*.

"Did you eat the cookie?" the mother asks.

"No," the toddler answers flatly – and unconvincingly.

The mother might find this scene adorable. Or she might grieve that perfidy has just been born in her child's soul. However, if this is the first time she has caught him in a lie, she should probably celebrate. The child has reached an important milestone in cognitive development.

By attempting to deceive his mother, the little miscreant reveals an emerging awareness: the ability to conceive of a mind separate from his own. He has demonstrated what psychologists call *theory of mind*: the ability to imagine another's thought processes, beliefs, and perspectives. When we do this, it is said we are *mentalizing*.

While it is possible to imagine a story where we don't have to figure out what others think, most storytelling and story-receiving requires theory of mind because most stories involve characters making decisions. When we watch a film, attend a play, or read a book we instantly and intuitively assess the inner lives of the characters in these stories. We evaluate their motivations, their reliability, their cultural background, their moral fiber – it is all part of the fun of hearing a story.

Why did humans develop the faculty of theory of mind? In our ancestral environment, social situations often required understanding the reasoning, knowledge, and motivations of others. Theory of mind enabled us to pick allies more competently and anticipate the actions of antagonists more accurately.

Back in the 1980s, Simon Baron-Cohen set out to identify theory of mind in children while working on his PhD thesis at the University of London.[65] It wasn't toddler fibbing he was interested in; rather, he wanted to test the hypothesis that mentalizing was a deficit in autistic children. However, coming up with a mentalizing test was surprisingly tricky. Imagine you're checking a three-year-old's ability to mentalize using a puppet theater scenario featuring Sally and Anne. The scenario unfolds as follows:

1. Sally and Anne are playing together.

2. Anne hides a marble in a basket for safe-keeping.

3. Both Sally and Anne leave the stage.

4. Sally returns to the stage and announces that she will look for the marble.

The experimenter's question: *Where do you think Sally will look for the marble?*

If the child answers "the basket," does this prove she has theory of mind? Not necessarily. The child knows the marble is in the basket so she does not need to consider Sally's perspective to answer correctly. To truly identify theory of mind, the experiment must test whether the child can recognize a false belief in another person's mind. Baron-Cohen developed a better scenario to achieve this:

1. Sally and Anne are playing together.

2. Anne hides a marble in a basket for safe-keeping.

3. Both Sally and Anne leave the stage.

4. Anne returns, takes the marble from the basket and places it in a box.

5. Anne leaves the stage again.

6. Sally returns to the stage and announces that she will look for the marble.

The experimenter's question: *Where do you think Sally will look for the marble?*

Most three-year-olds will assume Sally knows what they know; they will say that Sally will look for the marble in the box. But by the time they are five years old most children will answer that she will look in the basket, knowing that Sally will have a belief that is inaccurate.

So perhaps somewhere around 4 years of age we develop the ability to mentalize. Now, a parent who has curled up with a 3-year-old and a storybook might dispute that the child is unable to appreciate a story. Perhaps it is the pictures or the soothing brook of words that holds their attention; the ability to truly imagine another mind, to mentalize – as we do when we hear a story – appears to be a skill that develops relatively late in childhood.

This doesn't necessarily mean it emerged late in our evolutionary history. Experimental evidence suggests that great apes may, to some degree, share this ability with humans.[66]

The odd thing is that once we start developing theory of mind, we find it hard to stop, and we apply it to practically everything. In 1944, Fritz Heider and Marianne Simmel showed a number of subjects a simple stop-motion animation depicting the movement of geometric shapes. When they asked their subjects to describe the shapes, however, the subjects didn't talk about them as shapes.

The big triangle is a bully that is picking on the small triangle and circle, who are running scared but then figure out how to trick the big triangle and escape.

Or…

The big triangle is a jealous boyfriend of the female circle, and he is angry because he caught the circle flirting with the small triangle.[67]

We habitually apply theory of mind even when the object of our attention clearly lacks a human mind. Anyone who has owned a pet knows that much of the joy of having a dog or cat comes from imagining thoughts and feelings they couldn't possibly have. (For example, I imagine that my cat Hildy thinks our home is a hotel; she is constantly complaining about the sub-par quality of our service.)

Another example of the urge to mentalize comes not from the annals of psychology, but from the history of cinema. The early Russian filmmaker Lev Kuleshov interspersed identical shots of an expressionless actor with shots of a plate of soup, a girl in a coffin, and a woman on a couch. When he showed this edited film to audiences he was delighted at how they praised the actor's subtle performance. They reported that the actor seemed hungry, and then he seemed grieving, and then lustful – when really, it was all just the same shot reprinted in three different places.[68] We apparently don't need much evidence to develop theories of mind.

Not only do we have a compulsion to storify ourselves – we have an almost irresistible compulsion to storify others.

The Birth Of The Gods

Many scholars have suggested that the ability to mentalize is responsible for the advent of religion. One piece of supporting evidence for this hypothesis is that autistic individuals – who, as Simon Baron-Cohen established with his Sally-Anne experiment, often have a diminished capacity to mentalize – tend to show lower levels of religiosity than others.[69]

But how might theory of mind have led to religious belief? To explore this, let's imagine Shiki, our hunter-gatherer ancestor from the beginning of the book, tracking his prey in the wilderness.*

Shiki has subsisted on little more than leaves and roots for several days and hasn't had water since morning. He is physically weakened, yet his attention is singularly and obsessively focused on tracking a stag he has spotted. Shiki's eyes continually scan the landscape – he chooses carefully where to step on the ground to avoid making noise. He takes

---

* Here I am elaborating on a hypothesis by Andrew Newberg, whose work we will discuss again later in this book.

care to stay downwind of the stag, so his scent does not give him away. At this point, Shiki has no spiritual purpose, he merely wants to fill his gut so that he might avoid starvation. However, his intense concentration on the stag, almost mantra-like in its fixation, brings him into a trance-like state. His involuntary fast heightens this altered state of mind. Shiki's Atomist Self has been abandoned in an intense mental focus.

Finally, Shiki gets close enough to the stag for a clear shot. The stag is sitting peacefully on the ground. Sitting? After all this effort and tracking, why has the hunt suddenly become so easy? Shiki draws his bow, releasing an arrow that pierces the stag's heart. The animal dies within moments.

Shiki is euphoric! It's like the stag allowed himself to be killed. In this moment, Shiki has *mentalized* – not by imagining the thoughts of another human, but by projecting a mind onto his prey.

When Shiki returns to his tribe, he chooses not to brag about his hunting prowess, as his uncle Murrak sometimes does. Instead, he tells of a magical moment in the bush when the stag willingly sacrificed his life for the tribe.

"It was like the stag wanted me to kill him. Like he was making a sacrifice for me – for all of us!"

Shiki raises a piece of stag meat to the sky and he intones: "Thank you, brother stag, for this meat you have given us!"

The tribe is bewildered by this strange tale, but they like thinking that they live in a world with friends. They like the idea that hunting is not a cruel fight for survival, but a kind of cooperation. They like the idea their future might be cradled with such supernatural assistance. And of course, they are thankful for the meat. Maybe it is so.

That a scene such as this might have happened many

thousands of times in our ancestral history is hardly speculative. In hunter-gatherer societies the belief that there is a soul in animals, plants, and even inanimate objects is common. In their desire to understand and master their environment, early humans ascribed spirit, consciousness, and even intent to the elements that surrounded them: trees, animals, mountains, rivers.

Moreover, the idea that animals offer their lives to hunters has been identified by anthropologists as a recurring motif in hunter-gatherer mythologies. In these stories, the animal makes a sacrifice, and the hunter is obligated to respond with respect, gratitude, and careful treatment of the animal's remains to ensure that its spirit continues to cooperate with the tribe.

In our hypothetical conception, many of the important elements of religion exist in embryonic form:

1. The trance of the hunt leads to a heightened consciousness, resulting in feelings of unity.

2. The raising of the meat to the sky and the intonation of thanks represents an incipient ritual.

3. The story of the stag's sacrifice creates a nascent mythology with the potential to bind a people together.

East And West

So how was it that the spirit of the stag and his many elf and sprite cousins were run out of our world, to be replaced by a muscular grandpa sitting in the clouds?

It was probably a consequence of the rise of agriculture and animal husbandry. Hunter-gatherers could imbue their whole world with spirit – they could sacralize the environment that sustained them. But as soon as you till the ground, put some seeds in it and hope for rain, you also need to protect the land you've sweated over from the infringement of poachers. If you become dependent on the milk, meat, or labor of animals you have domesticated, you likewise need to maintain control of them and not let others purloin your bounty. Humans in this new economy needed to be able to own things – they required the concept of property.

It is awkward to own your gods – to in effect enslave them. To have a society with property, it is more comfortable to replace a spirit-filled world with a patriarchal figure that grants us the deed to parts of His creation. This God can grant us permission to claim and permission to rule – in the words of Genesis, to have "dominion." Humans had to de-sacralize their world to make space for their new technology and its new God.

In most of Asia, however, monotheism did not prosper. Why?

There are probably multiple drivers (including the amount of cultural contact with the Abrahamic strain of monotheism, which has been by far the hardiest), but an important factor may be the nature of growing rice – the dominant crop in Asia – as opposed to wheat, barley, or rye: the more common crops of Europe. To grow rice, you need to engineer controlled flooding. That means sharing water sources and building, maintaining, and administrating intricate irrigation systems that might include dams, canals, and sluices. Asian rice growers were forced to cluster together and depend on community-wide action; European wheat growers, on the other hand, might more easily imagine themselves as rugged individualists.

Perhaps this is why the east developed a religious and moral tradition that emphasized interdependence, rather than a sacred personage granting deed to individuals. Without the jealous One-and-Only God at its center, the east was able to retain remnants of animism much more than the west, with traditions such as Japan's Shinto, Mongolia's shamanism, and the myriad river, mountain, and local deities of Hinduism and Buddhist Southeast Asia.

Despite the Abrahamic God's jealousy, some traces of animism have managed to stow away in Western culture. The Catholic saints don't inhabit nature the way the animist spirits of the east do, but because of their life stories, they can have their particular concerns and interests. For example, St. Francis is seen as the patron saint of animals, St. Anthony as the patron saint of lost items, St. Patrick as the patron saint of Ireland. Though the Church teaches that saints have no power of their own, they can intercede with God on behalf of petitioners. Orthodox Christianity also venerates saints. Similarly, Sufi mysticism recognizes *Wali*, or friends of God, and Hasidic Judaism attributes intercessional powers to *tzaddikim* (righteous ones) and *rebbes* (spiritual leaders.)

However, to look to monotheistic religions for vestiges of animism is like looking for desserts in a recipe book for diabetics – not a crazy thing to do but also not the first place you would look. Forms of animism live and prosper everywhere in our modern world.

Children are natural animists. They delight in ascribing consciousness to inanimate objects, as do the stories that cater to children: *Goodnight Moon*, *The Little Engine That Could*, and many other picture books turn sundry objects into unforgettable characters. In Schoolhouse Rock's classic educational cartoon, *I'm Just a Bill*, a bill of U.S. law is turned into a character with aspirations and a personal history.

It's not just children that revel in animism. Personification happens in poetry too. John Keats addresses a Grecian Urn he sees in the British Museum thusly:

Thou still unravish'd bride of quietness,
Thou foster-child of silence and slow time

You might object that this is animism in only the loosest definition of the word. Older children, at least, don't think that a bill of law or a train engine has consciousness; it just tickles their imagination that it might. Likewise, Keats didn't think the urn was actually a bride-to-be or a foster-child. But I doubt that our hunter-gatherer forebears expected supernatural behavior from a tree; they didn't expect that its roots would become legs, and it would walk away to another place. Rather, they used their imaginations to understand and relate to the things around them.

I'll admit: I've seen the spirit in a tree first-hand. Between the ages of 8 and 13 I lived in Kenmore, a suburb of Brisbane, Australia. There was a park right off Dumbarton Drive where I used to go to play cricket or swim in the creek. On one corner of the park, right by a bend in the water, there was an enormous Moreton Bay Fig, a tree native to Australia, probably at least 50 yards tall. Much of my childhood was spent climbing this tree, hiding in its winding wall-like roots, or just enjoying its shade. It was a perennial character in the lives of my childhood friends and me: a giant, craggy uncle always ready to play, always willing to bless.

When I returned to visit Brisbane in my fifties, some four decades after I left, I naturally returned to visit the tree, which of course was still there. I walked around its broad trunk. I touched it reverently. It was like visiting a cherished relative.

Do you ever see the spirit in things?

Go to your wardrobe closet and open it up. I guarantee you will find souls there: lives you've led, poses you've struck, seductions and disappointments, sacrifices and indulgences.

Do you have a car? Does it not have a soul?

Have you been back to your childhood home lately? Do you not have a psychic blueprint for it that assigns a spirit to each hidden cranny?

Ruled By The Stars

Perhaps animism is sub-consciously practiced by some westerners. But the remnants of polytheism have a sturdier footprint here in another area: astrology is today something of a quasi-religion. A 2017 Pew Research Center survey found that 29% of Americans believe in it.

The origins of astrology are ancient – far older than Buddhism and the Abrahamic traditions. The ancient Babylonians divided the sky into the twelve signs of the zodiac. However, it was the Greeks that developed the idea of a natal chart, or horoscope, and it was the Greeks who associated their pantheon of Gods with individual planets. By linking a mythological system with celestial bodies, these ancients introduced a richness of association and personality to astrology that our moderns still milk.

Of course, astrology is not an exclusively western practice. Indeed, Vedic astrology has even earlier origins than the Babylonian variety, and today it remains tightly linked to Hindu culture and religion. Shakespeare has several allusions to astrology, usually tut-tut-ing:

The fault, dear Brutus, is not in our stars,

But in ourselves, that we are underlings.

Yes, underlings under stars. When we look up at the sky, we look for masters of our fate.

I have a couple of people in my life (whom, in fear of their reputation, I shall not identify) who enjoy astrology. Since I belong to the cult of science, I usually hear out their dissertations with bemused skepticism. How seriously do they take it? In terms of time and effort, quite seriously. They have downloaded software to track planetary movements; they read the various theorists and opine about planetary configurations; if they meet a new significant person in their life, they create natal charts for them.

How seriously do they take astrology in terms of its powers to divine the future? Well, much less seriously, but more than nothing. They have a low investment belief – would they be shocked if a particular divination didn't come true? Of course not. Deep down, I think they both know astrology's premonitory power is zilch, but they find it comforting and fun. And when I hear them speak about astrology at a moment that is high-stakes and high anxiety – frankly, I want to believe it too. I can't quite make the jump, but look: life is frightening. To be at the mercy of fate is frightening. When so many things – a medical diagnosis, a careless driver, or a weather event – can upend our lives, it's comforting to identify some sort of causation. What is interesting is that identifying causation is attractive to us *even when* intercession isn't possible, and *even when* the forecast isn't entirely sunny (although the output of this kind of divination is rarely dark.)

I've called astrology a quasi-religion because it lacks many of religion's defining elements, such as rituals or a moral system. Still, its appeal offers a clue to what draws many to religion: the need for an explanatory framework in a bewildering world. As humans, we need to find the causative links in story.

Devotion

What astrology lacks is the practice of devotion. If I were a sincere adherent of astrology, I wouldn't be called upon to light candles or say prayers to the constellation under which I was born. Oddly, the constellations have personalities, but they don't have personhood.

Just as a hunter creates the mind of his prey within his own, we can create another mind within ours – one that guides us through the world. We can cultivate a relationship with this mind that is mutually loving: we offer it praise, devotion, and gratitude, and in return, it offers us love and guidance. We can mentally re-parent ourselves.*

The practice of devotion is the nurturing of a bond rooted in mutual love, but it also requires humbling oneself before a deity, and acknowledging its power over us. Devotion is one of religion's most potent and time-tested tools for quieting the Atomist Self and strengthening the Connected Self.

Many non-believers view the requirement of faith as a weak point in monotheistic religions. I myself have admitted that I couldn't be a Christian because faith wasn't granted to me. But perhaps the requirement of faith is a boon rather a flaw.

To see why, let's imagine: what if faith weren't necessary? What if God appeared among us? I suspect that many Christians, Jews, and Muslims today are praying fervently for exactly such a moment, but I wonder whether it wouldn't wreck things.

---

* Christianity is most fond of the parental comparison (e.g., "Our Father, who art in heaven…"), but it is also present, with less emphasis, in Judaism. Islam, however, abjures such comparisons as they impinge on Allah's absoluteness. In Hinduism, Krishna describes himself in the Bhagavad Gita as "The father of this universe, the mother, the support, and the grandfather."

Imagine a scenario where God swoops down to some place where everyone is looking: maybe a Superbowl game or the grand finale of the Eurovision song contest. It turns out that we are indeed made in God's image, but the difference is that God is 14 feet tall. And very good looking.

At first, the TV producers don't know who this is, and they call security, but the special effects are so impressive they have to admit: this must be God. God makes clear that He doesn't want to spoil the show, so He takes a front-row seat, and the show goes on – with the participants more than slightly distracted. On the other hand, the audience at home mostly assumes this is an elaborate commercial tie-in.

The next day, however, the previous night's events are still unexplained. Once all human intervention is dismissed as a possible cause, the internet turns to speculating that maybe it's not actually God but an alien. Whatever or whoever it is, He keeps on taking a gleeful joy in proving that He's omnipotent and omniscient – the mind-readings of world leaders, the partings of the sea, the manna from heaven – at some point everyone just agrees to call a spade a spade.

This is partly because nobody wants to tick God off... because He really does seem eager to nurse a grudge. When He makes a point of going to Nietzche's grave, exhuming his bones, and making them dance to the chants of "Nietzche is dead, Nietzche is dead!" everyone must conclude that it is best not to be skeptical of His Divinity.

What is the point of this slightly blasphemous flight of fancy? It is to help us imagine what would be lost to us if God were a verifiably present and participating part of our world.

First, we would no longer have the mystery of faith. A God who is verifiably present would not demand the surrender of our rational mind. And if our rational mind is not surrendered, then our relationship to God becomes something transactional rather than sacred. God would become less like the mysterious divine and more like the beneficent, all-powerful superbeing we've just described. Worshiping Him (or Her) would not be a leap, but rather a transactional calculation – like humoring a benign dictator.

The searching nature of faith would also be lost. When David Brooks wrote about the dawning of his faith, he described how it stirred in him this way:

> *The word "faith" implies possession of something, whereas I experience faith as a yearning for something beautiful that I can sense but not fully grasp. For me faith is more about longing and thirsting than knowing and possessing.*[70]

Would the active, yearning nature of faith – its verb form – be lost if God were ostentatiously present in our lives? If every doctrinal and ethical issue were settled with finality, with God's incontrovertible say-so, would we lose some of the human dignity that comes from trying to discover His will?

For many believers, God is defined as much by absence as by presence. St. John of the Cross described his "dark night of the soul" as a time when God seemed distant, yet this absence matured and deepened his faith. If God were as accessible as our mobile phone, would we lose some of His or Her most vital lessons?

The idea that God has chosen to remain hidden is one that has been explored extensively by believers such Soren Kierkegaard, Blaise Pascal, and C.S. Lewis. For these thinkers, the absence of certainty about the existence of God is a gift to humanity from God.

For non-believing students of religion, our thought

experiment provides some hints as to how devotion does its magic.

Devotion is not merely the practice of loving and being loved, although it is certainly that.

It is the surrender of the rational mind to mystery.

It is the challenge of finding faith in uncertainty.

It is the search for a grander, more perfect mind.

The powerful thing about funneling your sense of the sacred through a personage is that it allows you to develop a relationship; if you think God is nature, as Spinoza did, it is probably more challenging to relate in a personal way to the divine. God becomes something of an abstraction, an everything-ness.

But, if you have a relationship with a person-God, then love can grow in that linkage. To believe that a deity knows you and loves you, and to return that love with humble adoration and devotion, is both comforting and empowering.

This is true for the monotheistic religions – Judaism, Christianity, Islam – but also true for the polytheistic traditions of Asia. In Hinduism, practitioners following the path of *bhakti*, or devotion, have a chosen deity (*ishta devata*) to whom they offer regular prayers and offerings. Or they may offer devotion to various deities, each representing a distinct principle or quality.

Even in Zen Buddhism, there are traces of devotion. We do a lot of bowing in Zen. We bow to the cushion, we bow to each other, and we do full prostrations to our teachers and the Buddha statue on the altar. If we ask our teachers what these deep bows mean, they will tell us that they are not bows to any deity nor are they any sort of worship, but rather a demonstration of respect to our teachings and tradition. Still, the fanatical care we take with the candles, the incense, the flowers, the food offerings on the altar – it's all more than a little reminiscent of the Hindu *bhakti* practices.

While we Zen Buddhists quiet our inner story with no story, adherents of devotional practices – whether they are polytheistic or monotheistic – counter their inner story with a bigger story than their own; a story that encapsulates them and contextualizes their life and existence, that urges them to surrender to and join with the divine.

The result is the same: to tame the Atomist Self and to approach a Connected Self.

CHAPTER 8

SYMBOLS IN THE MACHINE

Imagine a mind with all the capacities we have dedicated chapters to so far in this book: consciousness, self, emotion, memory, world-modeling, a theory of other minds – but as of yet, no language. A mute mind.

Add language to the mix and what happens? All of sudden, the experience of our organism becomes projectable, like an arrow. Infectious, like a virus. Transformative, like a spell. Language is the conductive wire of our culture.

Of course, there are other symbolic systems we could communicate with – the cave paintings at Lascaux seem to be telling a story of a hunt, for instance. Since those cave walls were painted, many a mime has amused (or annoyed) his audience with wordless tales. In our own era, Pixar's shorts have proven to be masterpieces of storytelling, despite containing no spoken language. Nonetheless, language is the richest symbolic modality we have, and it plays a role in nearly all our storytelling.

How did language arise? This question has inspired a brood of theories.[71] In the early 19th century, some thinkers speculated that language arose as an imitation of natural sounds; this came to be known as the *bow-wow theory*, giving an onomatopoeic name to the idea of onomatopoeia. Others guessed that perhaps it evolved from our instinctive cries or exclamations; this came to be known as the *pooh-pooh theory*.

In line with his theory of evolution, Charles Darwin emphasized a continuity between the simpler communication modes of animals and human language. In recent decades, this perspective has gained traction as the concept of human exceptionalism has correspondingly diminished. Darwin also proposed that gestures may have been a precursor to spoken language.

This idea of continuity was further developed in the late 20th century by scholars such as Michael Corballis and Gordon Hewes. They theorized that human bipedalism freed the hands for gestural communication, paving the way for a system of gestures to evolve. Eventually, spoken language took precedence because it did not rely on daylight or visual contact and was more efficient in terms of physical effort. This hypothesis has been bolstered by studies showing that similar parts of the brain are activated during gestural and spoken communication, hinting at an evolutionary link between the two.[72]

However, the true giant of language studies over the last century is undoubtedly Noam Chomsky. Chomsky proposed the concept of a Universal Grammar, an innate framework common to all human languages and hard-wired into the brain. According to him, this explains why toddlers can achieve the miraculous feat of language acquisition with such ease, even in the face of limited or fragmented inputs.

Initially, this led Chomsky to propose that the capacity for language emerged suddenly, around 100,000 years ago – a view that put him at odds with evolutionary scientists, who considered such a dramatic leap implausible. Over time, however, his position softened, and he acknowledged the possibility that language is not a single, unified process but rather a modular system, composed of parts that evolved much earlier for reasons unrelated to language.[73]

Large-Language Models

Today, with the rapid advancement of Large-Language Models (LLMs) such as ChatGPT, Claude, and Gemini, we are witnessing computers mimic the human miracle of verbal expression. While it's not guaranteed, these technologies may offer valuable insights into how our own minds work.

These AI models work by breaking words and parts of words down to tokens, and assigning different vectorial values to these tokens by absorbing vast amounts of text during their training – in effect, learning these tokens' meaning along a high number of contextual dimensions. For example, the word "cat" will be assigned different values depending on the words that come before it: *jazz* cat, *corporate fat* cat, *siamese* cat, and *siamese* will be assigned different values depending on whether it is followed by *cat*, *twin*, or *king*.[74]

Over time, the models learn to predict the next word in a sequence, enabling them to generate well-formed sentences.

These models can sometimes 'hallucinate,' producing outputs that are inaccurate or nonsensical. We can bemoan these faults, and people often do, but what seems to be far more noteworthy is their being as accurate as they are. Think of it: knowledge of the "real world" was not necessary to create a fairly reliable representation of the world; a deep probabilistic understanding of our culture's symbolic representations was sufficient. Intriguingly, these LLMs are sometimes more reliable than the material they are trained on. (Dario Amodei, the CEO of Anthropic, once suggested that this might be because all truth is connected, while untruths are not – if true, a comforting point.[75]) If humans are story animals, then LLMs are story machines.

There are both similarities and differences in how the human brain and LLMs work.

Computers use the binary language of 1s and 0s to do their work. Likewise, the human brain extracts complicated information from very simple signals: our brain's neurons either fire or do not fire, following an all-or-nothing principle. However, the human brain also includes a dimension that computers lack: neural plasticity. The brain constantly creates new neural connections, essentially re-wiring itself. These synaptic connections can have more or less strength, based on activity and neurochemical modulation.

Both the brain and LLMs utilize hierarchical layers of processing. In humans, this involves absorbing fine-grained sensory information – such as light signals from the eyes – and progressively extracting patterns, like edges, shapes, and eventually complex features, such as recognizing a face. As these signals ascend through neural pathways, they are processed by specialized regions of the brain for categorization and higher-level interpretation.

Similarly, in LLMs as well as image-recognition algorithms, data passes through multiple layers of artificial neurons. At each stage, the mathematical representations of tokens or pixels are progressively refined, enabling the model to identify patterns, contextual relationships, and ultimately generate meaningful and useful output.

It's easy to exaggerate the similarities between these hierarchical methodologies, however. A crucial difference is that the brain is not as predominantly feedforward as the computational models are. It contains rich feedback loops that support the predictive processes discussed in Chapter 6. Moreover, the brain engages in extensive parallel and distributed processing – capabilities that LLMs, as of yet, cannot match.

However, one of the most important breakthroughs in the development of LLMs was the introduction of the attention mechanism, which – as its name implies – analogizes with human experience. Computer scientists discovered that giving greater weights to the most important words or phrases in a given context allowed the model to prioritize relevant information while filtering out less critical details. This is analogous to how the human mind employs attention-focusing mechanisms, such as selective attention, to concentrate on the most pertinent stimuli in a given moment.

Can advances in computer science help us understand the human brain? As we've discussed, biological and technological systems can differ significantly. However, in both biological and technological evolution, similar innovations often arise along different paths because there may be a single most effective solution to a problem – a phenomenon known as *convergence*.

For instance, the wheel and writing systems were independently developed in multiple cultures. In biology, flight evolved separately in birds, bats, and insects, while echolocation arose in both bats and whales.

Similarly, parallels in innovation can occur between human invention and biological evolution. Cameras, for example, share many similarities with eyes: there are lenses in both, light-dimming irises in both, dark chambers in both, and photo-sensitive layers in both. It is quite possible, though not necessary, that biological and technological mechanisms tackling the same problems might come up with similar solutions. Perhaps advances in computer science will provide insights into the human mind.

Indeed, they already have. In his book, *The Alignment Problem*, Brian Christian recounts the story of a cross-disciplinary triumph: how computer scientists working on learning algorithms provided a vital insight that helped solve the puzzle of dopamine's function in the brain.[76]

AI has influenced the science and philosophy of consciousness by providing an ongoing proof-of-concept for materialist conceptions of mind. The idea that we need a spirit or ether to explain how our minds work has been in diminishing repute for many decades now, but as the output of generative AI becomes more impressive and human-like, the proposition that the wonder of the mind is the result of the human cells and their chemical and electrical processes – minus any additional hocus-pocus – grows in plausibility and hammers yet another nail into the coffin of Cartesian dualism.

The Meme Animal

Just as how signals compete for primacy in AI's microprocessors and in human neurons, signals compete for primacy in societies too.

Richard Dawkins coined the word "meme" in *The Selfish Gene*. Unfortunately, the word has been spoiled by people who use it to describe viral social media posts, but originally it was derived from the Greek *mimema*, meaning "that which is imitated." In Dawkins' usage, a meme can be any idea, concept or story that is passed on from one person to another by written, verbal or other means. He analogizes memes to genes, in the sense that they are both Replicators. And being Replicators, they are both subject to mutations and evolutions, with the memes that are fittest (meaning, likeliest to be shared) surviving, and the ones that are not fit (or not likely to be shared) perishing.

There are a few things to note about this idea.

First, it is an astonishingly powerful adaptation. Imagine a human tribe that relies solely on brute instinct to survive. They would hunt or gather food according to a hard-wired method or based on their lived experience.

Now imagine a human tribe that passes on not just its genes, but its memes. With each generation, its weapons for hunting become better. Its strategies for hunting become better. Its know-how for gathering and preparing roots and other plants become better. Eventually it learns about agriculture and animal husbandry and beer- and bread-making and – over many centuries – eventually acquires the skills and technology necessary to make our comfortable 21st-century lives possible.

Since we are talking about memes but this is a book about story, it might be useful to make a distinction: a story is a meme, but a meme is not necessarily a story.

We can also say that when we package up an idea or concept with a story, its fitness as a meme increases exponentially. Imagine this conversation:

"What is that?"

"It's a wheel."

"Huh? It's just a round thing."

"We can use it for transportation. It will change the world."

"Ha-ha."

Every solution needs a story.

"What is that?"

"Oh, it's a wheel. I was cutting a tree the other day and after it fell it rolled down the hill all by itself. I thought to myself, maybe this rolling motion might help us move things easier – instead of carrying them or dragging them, we can roll things places."

"Wow!"

Finally, we can also say that memes are as selfish as Genes are. They don't care (metaphorically speaking*) about the furtherance of the human species, only the furtherance of themselves. As Dawkins conceived it, the memes that are useful to humans do tend to be passed on more than those that are not useful. That's still generally true, but anyone with even a passing acquaintance with social media knows that the memes that replicate there are the ones that provoke the strongest emotions, creating certain social and political dysfunctions.

Monkey See, Monkey Do

How did humans become capable of adopting and passing on memes? An important clue to answering this question came in the early 1990s, when an Italian neurophysiologist named Giacomo Rizzolatti was conducting experiments on macaque monkeys at the University of Parma, in Italy. As with the discovery of the DMN in St. Louis many years later, this breakthrough came almost completely by accident, but smart and alert researchers were required to figure out what was going on.

Rizolatti and his colleagues wanted to find out what parts of these monkeys' brains were responsible for what physical

---

* This common formulation is more proof of how difficult it is to avoid mentalizing!

actions, so they attached electrodes to single neurons that recorded when certain brain cells fired.

They had identified a neuron that fired when a monkey grasped at an object. Fantastic! A job well done. But then on a break, when a researcher grasped the object in the same way and the monkey saw it, the same neuron fired again.

How could that be? This single neuron was part of a network that controlled a physical action, but it was also part of a network that recognized that physical action in others? The team tried the experiment again, and again got the same results.

What these researchers had discovered was mirror neurons. These are a type of brain cell found in humans and other animals as well, but primarily the primates. Because these cells fire both when a subject *sees* an act and when they *do* the act, mirror neurons are thought to facilitate imitation.[77]

The Parma team's discovery unleashed a torrent of research, and we now know quite a bit more about this topic. For example, mirror neurons do not need to be triggered just by sight; they can also be triggered by auditory cues or even our imagination.

Moreover, mirror neurons appear to be sensitive to the *objective* of the action. Leonardo Fogassi, one of the original researchers at Parma, later found that certain neurons in a macaque's brain fired when the monkey saw food being grasped to eat, but not when it was grasped to be placed in a container. It seemed the brain was interpreting the *purpose* behind the action and associating it with the specific behavior.[78]

Mirror neurons aren't just about imitation; they also seem to be associated with empathy. Mirror neurons have been observed to fire in a sympathetic way when we see others in pain, or when we observe a flood of emotion in others, causing a simulacrum of those feelings and emotions in us. The neuroscientist V.S. Ramachadran comments:

> *It is difficult to overstate the importance of understanding mirror neurons and their function. They may well be central to social learning, imitation, and the cultural transmission of skills and attitudes – perhaps even of the pressed-together sound clusters we call "words." By hyper-developing the mirror-neuron system, evolution in effect turned culture into the new genome.*[79]

In effect, our brains provide us with specialized neurons that act like meme-o-graph machines, absorbing and recreating the culture we see around us.

Lifespan

Our brains are not the only part of our organism to be optimized for the passing on of culture. Our lifespan itself may have contributed to the rise of culture.

Humans are dependent on their parents at least until their early teens; no other animal has an infancy that is quite as long. This period of reliance creates an opportunity for apprenticeship that allows ample time for the absorption of modes and methods.

On the other end of life, only whales have as long a life after menopause as humans do. You would think that from an evolutionary point of view, the extension of life after the end of fertility would be wasted, but theorists believe that this allows grandmothers to assist in the upbringing of their children – and thus, pass on know-how.[80]

It's not just grandmothers that contribute; we might have

evolved to cash in on grandfather wisdom, too. The human lifespan of both sexes is much more extended than in most other animals, including our great ape brethren. Perhaps this is an adaptation that helps us preserve ways and memory.

Victor Of Aveyron

Is it possible to imagine ourselves without culture? Our culture has imagined it. In mythology, the founders of Rome were Remus and Romulus, two abandoned children adopted by wolves. According to legend, after growing to adulthood they argued over where to establish their city, and Romulus killed Remus to settle the disagreement – which, I suppose, is why the Vatican isn't in Reme.

True life accounts of feral children do exist, however.* One of the most famous is Victor of Aveyron, the inspiration for Francois Truffaut's film, *The Wild Child*. Victor was discovered in 1800 at around age 12, wandering the woods near Saint-Sernin-sur-Rance. He was found naked, with matted hair, feet toughened into leather-like hardness, and his body covered in scars and scratches. Victor likely survived by foraging for berries and roots, or perhaps by stealing from nearby farms and gardens. He had no ability to speak and could communicate only through grunts and cries.

This true-life tragedy did not come with a magical redemption at its end: sadly, Victor was never able to integrate into society very well.† As is my wont, however, I will use yet

---

* A famous recent one was the story of Oxana Malaya, a Ukrainian child abandoned by her alcoholic parents in the 1980s and taken in by some dogs who lived in a nearby shed. She eventually adopted dog-like behaviors, moving around on all-fours, barking, and eating directly with her mouth.
† In his film, Truffaut – to his credit – does not sugar-coat this outcome.

another sad case as a springboard for an exercise in imagination. What if you and I were stripped of the culture passed on to us? What if you and I were Victors – culture-less forest hermits?

One of the ways we define ourselves is by our traits: whether we are intellectual or not, social or not, artistic or not, polite or not, funny or not. If we were Victors, none of these dimensions of personality would have meaning for us. That is partly because these terms only make sense in comparison to others. Just as how you wouldn't need a thermometer if every day had the same temperature, you wouldn't need differentiating character terms when your whole world is just you.

However, it's not just that the concept of character would lose its *relief*, it would also lose its *function*, because as we discussed in an earlier chapter, the function of the self is to act as a bridge to society. In hermitage, the self makes less sense, which is likely why solitude has been used as a tool for spiritual advancement in many traditions.

Let's set aside Victor's social isolation for a moment to consider his situation more fully. Imagine that a tribe of Victors joined together. Even with the amazing Language Acquisition Device that Chomsky identified, it seems unlikely they would develop a fully formed language. A group of Victors living together for years might, however, develop a proto-language, using gestures or sounds. Perhaps it would consist of a dozen or so nouns to identify berries or roots and a few verbs to describe actions like sleeping or walking.

Such a tribe would need many centuries – or even millennia – of cultural development to develop the higher capacities that Hamlet was so proud of:

What a piece of work is a man! How noble in reason! How infinite in faculty!

The Palace Of Indra

Reader, I don't mean to flatter you, but as you sit there processing this stream of symbols, you do seem infinite in faculty to me. But it's not just *your* doing. And it certainly isn't mine.

Back in 2010, researchers at Princeton University used fMRI to record the brain activity of a subject telling an unrehearsed story. They then scanned the brains of listeners as they heard the same story through headphones. Remarkably, the listeners' neural activity closely mirrored that of the speaker. Storytelling brought them all into a mental synchronicity.[81]

Perhaps there is some synchronicity between us as I write and you read, but honestly, I hope not too much. My favorite books are the ones where I read half a page, then stare off into the distance for a few minutes, turning over the ideas that have been invited in. I hope to have this effect on you: that you see more than I, disagree in part with a "yes, but..." or perhaps just shake your head sadly at my meager understanding.

But the only reason we can do this together is because we share a language and share a common repository of knowledge and culture. The tribe of Victors doesn't know what a book is, much less how to decode and process these words and knock them about in their head to create new syntheses.

Writing this book after the advent of generative AI, it seems obvious that *Creation* is really just *Curation*. Generative AI can create texts and images that we would undoubtedly credit as creative if they came from humans, and they are doing it by absorbing, processing, and rearranging the symbolic elements of our culture.

I am doing the same thing as I write this book. I am weaving together the theories, discoveries, and reasoning of millions of brighter lights that came before me. I say millions because each source has been nourished by thousands of uphill tributaries, all the way back to the Buddha and beyond, because he himself was nourished by the rich Hindu tradition. I suspect that even the handful of musings in this book that felt original as I conceived them have probably been arrived at independently by dozens or hundreds or thousands of others before me.

As history has advanced, our interconnections have become faster, more plentiful, and more complete. LLMs represent the apex of this process, but they are also a metaphor for what human culture has always been. Our culture is a vast neural network like LLMs are, and we are all neurons. We make many synaptic connections with books, videos, social media, conversations. The memes/stories enter us and are passed on by us.

Strip yourself from your place in this grand mind, and what is left? Compare the *You* that is reading this to *You* the Victor – grunting, naked, mind-darkened. Could you even call that person *You,* even if he or she carried every single quirk and proclivity encoded in your once-in-an-eternity DNA?

In Buddhist mythology, the great god Indra lives in a luxurious palace on Mount Meru. One of the rooms is spanned by a beautiful net that stretches infinitely in all ten directions. At every intersection in the net, a brilliant jewel is suspended, each reflecting and refracting the light and image of every other jewel in the infinite web.[82]

Each one of us is a jewel in Indra's net. Our selves are so definingly interdependent with each other that existing in isolation is almost inconceivable.

CHAPTER 9

THE BIRTH OF MYTH

I magine the human capabilities we've described in previous chapters – consciousness, self, emotion, episodic memory, mental modelling, theory of mind, language – as tumblers in a lock. Over the expanse of evolutionary time they are developed, each to respond to an evolutionary need, and when they are finally aligned together this grand mechanism can turn and the lock will spring open; from the vault of human experience, a geyser of narrative will burst out.

These stories will shape our selves. They will shape our culture.

So when did this fusion of capabilities occur? Tracking the storytelling proficiency of pre-literate people is challenging, but it's likely we developed this ability early in our evolution. Archaeologists have discovered neanderthal graves sites where the dead are buried with their personal tools and belongings.[83] Was this done to prepare them for the afterlife? If so, then our hominid cousins must have had the capacity to create myth. Similarly, the graves of homo sapiens buried more than 30,000 years ago have included possible offerings and ritual instruments that hint at symbolic thought.

Turning to storytelling artifacts, the Lascaux cave

paintings in France are among the earliest discovered. These paintings, created around 17,000 years ago, are widely celebrated for their creativity and craftsmanship: the images are both painted and engraved, and they use the natural curves of the cave walls for effect. A smorgasbord of competing theories has been offered to explain these images, which seem to depict scenes from a hunt. Some details are so strange and specific that they appear to be the handiwork of a prehistoric David Lynch: a wounded bull, with its entrails spilling out, stands beside a reclining man with an erection, while a bird perches on a stick nearby. Clearly, some sort of narrative is at play here, along with a great deal of symbolism.

Beyond image-making, writing has arisen independently in multiple places throughout history. Written language has often had a common mode of gestation. It frequently begins as an accounting procedure for recording the quantities of physical goods, then acquires more detail and sophistication until it comes to include a record of more than just the items and their numbers.* The earliest instances of proto-writing appear in Mesopotamia around 8,000 BCE, where they used clay tokens to represent varieties and quantities of goods. To avoid the tampering or rearrangement of these tokens, the Mesopotamians contained them in hollow pottery balls, called bullae. They would sometimes make an imprint of the inner tokens on the outside shell of the bullae so it would not need to be broken to reveal its contents. This practice initiated a shift to a different modality of writing, the pressing of symbols into clay, which became the form of writing we call cuneiform. By

---

* Despite their other sophistications, the Incas did not develop a writing system. Perhaps this was a tragic case of path dependency. They kept their accounts in *quipus*, systems of knots, that were not prone to easy rearrangement as characters were.

2,000 BCE writing systems had been developed in Mesopotamia, Egypt, and the Indus valley, and by 1,000 BCE they were devised in China and Mexico too.[84]

By the time the Sumerian epic of *Gilgamesh* was written, around 2,000 BCE, many of the staple themes of myth were set: the intervention of Gods into human affairs, mortality, nature, civilization, the pursuit of a meaningful life, hubris, and humility.

Our definition of story is broader than simply myth, however. Stories of influence have had an awesome effect on our world. They have built skyscrapers and raised planes to the sky – as well as put our planet into climate peril. They have connected information around the Globe and placed us at the dawn of a new era: beyond the evolution of DNA, beyond the evolution of culture, to the new and lightning-quick evolution of learning algorithms.

Today, our lives are saturated with stories. At work, we hear and tell business stories. When we get home, we share the stories of our day with our loved ones. Then on a television or digital device, we will soak up stories to entertain us, go to bed and dream stories, then wake up and start the cycle again. Stories fill our days, binding together a civilization that is truly global.

Yet, as stories have expanded our world, they have shrunk our hearts. They have made us believe we are unitary, independent, persistent: just a story – and an oftentimes disappointing one.

To the revisit the metaphor that gave this book its title:

The Arrow has pierced our soul, and our soul has taken the Arrow's form.

Since the dawn of time, since Gilgamesh and before, we have turned to myth to cure our lacking inner story with a transcendent outer story. The institutions we call religion have preserved and developed these myths, combining them with rituals, systems of ethics, and communities united in sacred purpose. Religion has been a part of human life for as long as we can trace, but in the last couple of centuries these institutions have come under a climate of criticism.

The Atheist Critique

Many thinkers have aimed their rhetorical guns at religion, but much of the siege has occurred in silence. Modern society operates under values and assumptions that have not been kind to religion. It is worth noting that Buddhism, particularly in its Western manifestations, has managed to evade much of this scrutiny. An atheist intellectual like Sam Harris, for instance, can be antagonistic to deism while remaining friendly toward mindfulness meditation.

The atheists' critique of religion often begins with disputing its cosmological claims. A hero of these pages, Charles Darwin, is part of the reason the truth claims of religion have been cast into doubt. However, the scientific method itself is primarily responsible for this breach. When we are habituated to crediting as truth only that which can be verified experimentally, a postulate like God's existence – unverifiable by such standards – inevitably suffers. It is true that God's existence is unfalsifiable as well unprovable, but that is not a point of epistemic strength.

Turning to religion's ethical function, while atheists may grudgingly acknowledge its role in maintaining societal mores, they are quick to highlight historical episodes where religious belief led to reprehensible outcomes: the Crusades, the

Inquisition, and so on. Many religions harbor tribalist tendencies that have produced abhorrent actions against out-groups. Buddhism has also shamed itself, with the persecution of the Rohingya in Burma and other similar instances in Sri Lanka. The facile way to react to this would be to point out that atheists have been responsible for loathsome outcomes, too – Mao, Stalin, and Pol Pot, for starters. There is some justice to this riposte – the worst moments of religion should not be associated unfairly to its totality – but still, it would be a mistake to ignore this criticism all together.

Yet another critique of religion is that its ethical systems often fail to evolve with the times. Secular ethical frameworks, by contrast, can adapt more nimbly and require no religious basis to gain acceptance. For example, it has been over a century since women gained the right to vote in most Western countries, yet the Roman Catholic Church still refuses to ordain women as priests. From the Church's perspective, this rigidity is understandable. If you are claiming that your mores and practices are the preferences of an Eternal Being, then your eternal truths should not change jauntily with the cultural weather. It would alienate the constituencies that have built their lives around them. The Second Vatican Council took place six decades ago, and still today its mainly procedural changes are grieved by some in the Catholic Church. More sweeping reforms, such as gender parity or recognition of same-sex marriage, will likely provoke significant conflict and ensuing heartache.

Unfortunately, as more forward-looking people perceive their religious homes as out of step with their values and leave, a "rump effect" occurs: the remaining constituency becomes, on average, more conservative and resistant to reform. This can become a death spiral: the more a religion needs change, the less able it is to effectuate it.

For those who have left their childhood spiritual homes, religion may seem more focused on stifling natural impulses than on guiding adherents toward sacred experiences of the world. Perhaps when religious institutions feel besieged, focus on their core spiritual function is lost.

Once atheists have detailed religion's flaws, they are often challenged to explain its historical ubiquity. If religion is so devoid of value, why has it been so pervasive? The atheist might reply that religion is a parasitic meme – an idea capable of replicating itself despite its lack of utilitarian value. Like a chain letter that promises grand rewards if obeyed and dire consequences if ignored, religion's virality, they might argue, is proof not of its worth but of its believers' gullibility.

However, it is on these pragmatic grounds that the atheists' critique is most vulnerable. While research results vary by cultural context, large-scale studies frequently show a correlation between religious practice and increased happiness, as well as higher life satisfaction metrics. There is even stronger evidence linking religiosity to social indicators like lower divorce, lower crime, and better health. If religion "works," should we dismiss it solely because we cannot validate its truth claims? Many medicines on the market have unknown mechanisms of action, yet we use them without controversy because they have been verified as effective. To reject religion because its truth claims are unproven would seem profoundly irrational.

Still, rightly or wrongly, "it works" is not the verdict modern Western society has rendered on religion.

The State Of Religion

When Gallup asked Americans in 1958 whether they had

attended a place of worship in the past seven days, 49% said yes. By 2023, that number had nearly halved to 26%. In many European countries, that statistic is even lower, often in the teens or single digits. All forms of religious practice suffered a significant decline after the pandemic, as many practitioners did not return to in-person services.

This decline is not analogous to the fall of newspapers or linear television. Media networks and print periodicals have been with us for only a few decades, but religious institutions have been central to human culture for millennia. In fact, some form of religious life has been a cross-cultural constant in all human societies for as long as we can tell. This part of being human has taken a terrible beating in the last century or so. It is a development of exceptional historical salience.

As religion falters, the cultural gatekeepers and tribunes of modern society have largely greeted its existential crisis with indifference. Their sympathies often lie with secular values, and they neither fully understand nor empathize with the religious impulse.

Sadly, we are in greater need of religion today than ever before. Our hyper-individualistic culture has put the Atomist Self on a diet of steroids. As a result, we are more depressed, disconnected, anxious, and aimless than ever before. We need the techniques and traditions of religion to help us navigate our modern culture, but our modern culture is at odds with the very tool required to deal with it.

I write this not as a personal prescription to you. If you do not practice religion and still find meaning and satisfaction in life, then more power to you! If you do not need religion, then you do not need it.

Still, many of us do need religion – whether in a deist or non-deist form. Among those living with a sense of dissatisfaction or incompleteness, some may not even sense that a religious practice can offer restoration. The rest of this chapter is an attempt to imagine a religious culture that might co-exist fruitfully with modernity.

Multi-Modality

Why is this book, with its flag of sympathy planted firmly in the Buddhist camp, concerning itself with religion in general? Many Zen Buddhists would reject their practice being placed in that category (or any category.) It would probably make for a more attractive offering to position Zen Buddhism as a non-religious form of spirituality.

Indeed, part of the reason why Zen Buddhism appeals to me is because it sidesteps some of the critiques of religion previously mentioned. Zen Buddhism contains none of the cosmological dissonance of the deist religions. Buddhism's moral system, the Precepts, is apt for adaptation with the times. I prize that Zen is suited for modernity.

However, despite this, I have widened the topic to religion in general because I do not believe that a meditation-centered, non-theist, non-devotional practice is apt for everyone. A healthy religious culture in modern society needs to be multi-modal.

When I reflect on the sanghas I have been a part of, I see a recurring pattern: academics, artists, and the helping professions are over-represented; intellectuals and introverts are over-represented. We are far from a cross-section of society, and that makes me suspect that perhaps we are not for everyone (which is not to say that we should be.)

Human temperaments are so diverse that perhaps there is no single indicated remedy to their original malady. In India,

Hinduism offers a pluralistic array of paths and practices for spiritual growth, tailored to individual temperaments and circumstances. *Raja Yoga*, sometimes called the royal path, is the meditative yoga. Perhaps it is a reasonable analog for Zen Buddhism.

But Hinduism also offers other paths: *Bhakti Yoga* is the path of devotion: you practice it by developing a loving relationship with a chosen deity. Despite the fact that this deity is selected from a pantheon of options, Bhakti Yoga might be comparable to the devotional practices of Christianity, Islam, and Judaism.*

Karma yoga is the path of action; you practice it by bringing awareness to your acts of service to others. You can practice Karma Yoga by being a devoted parent, teacher, or performing any social duty, really. Any eastern or western tradition that incorporates activism, service or charity has some Karma yoga in it.

Jnana Yoga is the path of knowledge: you practice it by studying not just books but the nature of reality and the self. This more intellectual approach may be akin to Talmudic scholarship or the Jesuit tradition – or the aspirations of this book.

Perhaps in a westernized version of the Hindu paradigm, Zen Buddhism might be seen as the best path for some, but not for all. The idea that religious practice should be fitted to temperament is not alien to the west: William James recognized it in *Varieties of Religious Experience*.

Perhaps in such a pluralistic religious culture we can make individual choices without either suffering or imposing

---

* It's worth noting that deist religions often contain contemplative and mystical traditions within them, such as Kabbalah in Judaism and Sufism in Islam. These traditions might be more akin to Raja yoga.

chauvinisms or tribalisms. We could recognize the religious impulse as a universal and human phenomenon. We could respect diverse forms of faith and religious practice as attempts at service and self-improvement, but equally respect the choice to not believe or not practice.

Traditional And Reformist

I've mentioned that mores can't change easily when they are the preferences of an eternal being, but necessary conservatism isn't just an issue for the deist religions. In Zen Buddhism, we face the task of adapting a Japanese monastic tradition to a modern, lay audience spread across many parts of the world. We have a lot of figuring out to do to make it work, but as we do, how do we decide what to preserve and what to change?

When I first started sitting at the Zendo, the robes some practitioners wore felt at best a bit too formal, and at worst a kind of affectation. When I went on retreat for the first time, robes were required, and it was there – when I had a chance to try them myself – that I came to appreciate them. They minimized distractions from bright clothes or comely limbs. They seemed to unify the sangha in the same color of black. They were comfortable to sit in. And most of all, putting on the robe helped put you in mind for practice.

If I were asked what to change in the rituals of the Zendo, I might surely have some leanings as to what "clicks" for me most. But they would not be convictions, because I've seen that my leanings tend to change as I learn and explore the ways of Zen, just as how my feelings about robes have changed. Moreover, I understand that what clicks for me might not click for everyone, so I would be frightened to change anything lest I destroy centuries of tradition and learning by an ignorant whim. If temperaments for change span from the radicalism of Robespierre to the conservatism of Burke, I am Robespierre for most things – a little *too* in love with the new, to be truthful – but a very cautious Burkean when it comes to my religious practice.

Conservatism, when it comes to ritual and procedure, seems safer than rash tinkering, but when it comes to ethics, reform is urgently needed in most religions. However, it should be a reform designed not to take us to a new ethics, but to return us to the spirit of the original teachings. There is little in the Ten Commandments that offends modern sensibilities. Similarly, a close reading of Jesus's sayings in the Testaments reveals very little that would trouble a progressive audience. In fact, Jesus's frequent calls to care for the poor and marginalized would likely garner applause from many left-leaning seculars. The over-arching theme of the Gospels, the one that jumps from every page, is to love and care for one another. In contrast, the modern ethical friction points – such as homosexuality, pre-marital sex, gender parity, and abortion – receive far, far less attention.

This is not to say that churches, synagogues, and mosques should abandon their specific moral teachings altogether by... endorsing libertine sex, for instance. But an emphasis on the important principles and a measure of doctrinal humility might gain more ethical compliance than high-handed prescriptivism unmoored from the original spirit of the scriptures.

Communal But Not Tribal

While religion has suffered a slow bleed over the last century, one option from the religious affiliation pull-down menu has prospered: "spiritual but not religious," or SBNR as some have abbreviated it. This category is large partially because it is so vague: it might include a belief in a deity, or some fuzzy pantheistic feeling, or a preference for something more eastern and new-agey. It may even include people without any particular belief, who shy away from labels like "atheist" or "agnostic" because they seem hyper-rational or unfashionably unfeeling. In any event, SBNRs acknowledge the essence of religious stirring while rejecting the institutions that come with it.

In some ways, SBNR is a natural extension of the individualist trajectory religion has taken in the centuries since Luther nailed a paper to a church door. But disconnecting completely from institutions comes at a cost.

A friend of mine was a member of the 82nd Airborne division, the U.S. army's elite parachuting infantry. Commenting about the army's training, he once told me: "We know how to make soldiers. We've been doing it since Roman times, so we've gotten it down pretty good." I will acknowledge that it is a little perverse to compare the training of a warrior to the training of a religious adherent, but perhaps it's not so apposite. Both involve the transformation of a psychology from the self-centered to the other-centered. Religions have been using ritual, community, pastoring, and liturgy to do this for centuries: they know what works. To throw away this tradition and learnings would be foolhardy.

To be Spiritual but Not Religious is like identifying as a golfer, but not wanting to take golf lessons – not wanting to be taught first-hand what others have discovered about the game. It's not ever knowing the camaraderie of a foursome. It's not enjoying the pastoral beauty of a golf course. To be Spiritual but Not Religious is to take hacks by yourself in a driving range, behind the mall. If you want to be a driving range hacker instead of a golfer, you can make that choice, but you have opted out of a fuller experience. If diluted enough, any potion just becomes tainted water.

Of all the elements of religion that SBNR rejects, the most soul-damaging relinquishment is that of community. Religious communities can be like extended families that shield and protect us from crises and misfortune. They can be a workshop for good deeds. They can be a fountain of connection and a salve for loneliness.

Indeed, turning away from community is the defining characteristic of SBNR. Consider a thought experiment: imagine that a gaggle of SBNRs begins attending a Unitarian-Universalist church. Their wide range of beliefs and non-beliefs would not change an iota; it would be accepted by the UU congregation as such, since this movement is not centered on any belief system, and welcomes believer and unbeliever alike. Would you still call these people non-religious if they were actively engaged in a church community?

Community is a definer of religious activity. In many ways, community is the very point of religion.

However, community also has a dark side. Identification with the in-group can sometimes lead to a distancing from – or even hostility toward – out-groups. It is not a mystery why this is. In humans, the tribal instinct is powerful. Tribal belonging often intensifies through differentiation from others; it's not enough to be part of a community – it must also feel like the best community.

This dynamic partly explains why religion, at times, has been responsible for terrible events. When tribal fervor overtakes the core values of kindness and compassion, religion's sacred spirit is eclipsed by its more base instincts. Humanity's worst deeds have often been carried out in the name of righteousness. In today's diverse, multicultural societies, this antagonistic tribalism is increasingly incompatible with social harmony – and it must be said, also runs counter to the true and original sacred teachings of every major religion.

For religion to thrive in modern society it must be rooted in community while resisting tribal exclusivity and animosity toward out-groups.

Interspirituality

A key aspect of a religious community's attitude to out-groups is how it views the Gods, practices, and beliefs of others.

Not to seem cynical, but for religions this poses a marketing dilemma. On one hand, as with any form of marketing, product differentiation is critical. If your God is not the one true God, and your Way is not the one true way, why should anyone choose it? On the other hand, if your God offers paradise to believers while abandoning millions who have never heard His name, it risks making your God seem petty and small. In a multicultural society it is difficult to credibly tell a Christian family that their extraordinarily kind Muslim neighbors are destined for hell. Nevertheless, some denominations persist in making this claim – though their numbers are dwindling.

In our diverse societies, fostering a framework for the respectful co-existence of religions not only promotes harmony but also enhances spiritual credibility. Looking to other religions for inspiration can be a source of enrichment and strength. The goal is not to merge or homogenize traditions but to cross-pollinate them, uncovering shared human truths across cultures. This practice is called interspirituality.

Religious traditions can be placed on a spectrum of interspirituality. At one end of the spectrum are religions that view those outside their group as apostates who must be suppressed, converted, expelled, or even eliminated. This represents the most extreme and exclusionary stance.

Next on the spectrum are religions that promote self-isolation, discouraging their followers from interacting or engaging socially with those outside their group. Examples of this include some conservative Jewish communities, the Amish in Pennsylvania, and certain other fundamentalist Christian sects.

Then there are those religions that consider the outgroup in mortal error – that because they do not believe in the "right" God they will be doomed to perdition. This was the doctrine of the Catholic Church before Vatican II: there was no salvation outside the Church. After Vatican II, the doctrine shifted to acknowledge that people of other faiths could achieve salvation if they sincerely sought God.

Somewhere in the middle are the practices that show respect for other traditions, but little curiosity. The other religions might not be in mortal error, but they are not the true path and we don't have anything to learn from them.

As we cross to the other end of the spectrum, we find religions that purposefully look beyond their perimeters for inspiration. One of the trailblazers, in this regard, was the Bahá'í faith, founded by a man who would become known as Bahá'u'lláh. His story is worth telling, so let's pause and do that.[85]

Bahá'u'lláh was born Ḥusayn-'Alí Núrí, in 1817, to an aristocratic family in Tehran, and was raised as a Muslim. In his twenties, he became a follower of the Báb (meaning "the Gate"), a splinter figure from the Shia tradition who claimed to be a divine manifestation on par with Jesus or Muhammad. The Báb's teachings challenged both religious and political authorities. He advocated for reforms such as the equality of women, the modernization of education, and the abrogation of outdated Islamic laws. Like John the Baptist in Christianity, the Báb prophesied the arrival of "he whom God will make manifest," a figure greater than himself.

The Báb's liberalizing message was deemed deeply threatening by the Persian authorities. He was arrested and executed by a firing squad of 720 rifles. According to legend, he survived the first volley but was killed by the second.

Following the Báb's death, Bahá'u'lláh emerged as a prominent leader of the Bábi movement. He narrowly escaped execution himself and was imprisoned in a notorious dungeon, where he claimed to have received his first divine revelations. Even in captivity, he was seen as a threat to the established order and was exiled to Baghdad. Many Bábi followers joined him there.

After a conflict with his brother Yahya, who sought to lead the movement despite lacking Bahá'u'lláh's spiritual gifts, Bahá'u'lláh retreated to the mountains of Kurdistan. For two years he lived as a dervish hermit, until he was persuaded by his followers to return to Baghdad and rebuild the Bábi community. His growing influence alarmed the Persian regime, which urged the Ottoman authorities to extradite him. While the Ottomans refused, they cajoled Bahá'u'lláh to move further from Persia, first to Constantinople, and later to Acre, where he declared himself "he whom God will make manifest."

Bahá'u'lláh expanded on the Báb's idea of progressive revelation – the belief that throughout history, God had sent a series of messengers to the world: Adam, Noah, Abraham, Moses, Zoroaster, Krishna, Buddha, Jesus Christ, Muhammad, the Báb, and finally Bahá'u'lláh himself. While God is eternal and unchanging, His teachings evolve to suit the cultural and historical context of each age.

It's worth noting that this inclusivity doesn't come at the price of sacrificing a claim to specialness: the teachings of Bahá'u'lláh are still the latest and greatest. So we have an elegant solution to the marketing dilemma aforementioned! Bahá'í believers today maintain a curiosity and openness toward other traditions, embodying the spirit of interspirituality.

There was another off-shoot of Islam that pushed the boundaries of interspirituality: it is what some have called the Western or Universal Sufism of Hazrat Inayat Khan[86].

Inayat Khan was born in 1882 to an aristocratic and artistic family in Gujarat, India. From an early age he was a musical and intellectual prodigy, publishing his first book on music at the age of fourteen, and gaining recognition for his mastery of the Veena, a stringed instrument. For Inayat Khan, music was intimately linked to spirituality, so his pursuit of the former inevitably led to the pursuit of the latter. A practicing Muslim, he took a Murshid, or teacher, to teach him the ways of Sufism.

Sufism is not a sect of Islam like Sunni or Shia. Sufism is the mystical practice of Islam, a layer that is placed over the usual practices and obligations to bring a more immediate experience of God. The etymology of the word "sufi" is controverted, but one possibility is that it comes from the word taste – it is said that Sufis don't merely want to worship God, they want to taste Him.

So when Inayat Khan went to the west to play music and spread sufism, he saw his religious practice as a sauce on top rather than the main dish. That the Americans and Europeans he met already practiced Christianity or Judaism didn't ruffle him; rather than try to win converts with proselytization, he saw what he did as a complement to their traditions. He taught Sufi techniques of music, movement, chanting, breath, and meditation to these westerners, assuming that their God was his

God.

One of the rituals he introduced was the Universal Worship Service that recognized the wisdom teachings from many traditions.

Inayat Khan's teachings are carried on today by his grandson, Zia Inayat-Khan and the Inayati Order.

If you read Inayat Khan's very beautiful Ten Sufi Thoughts you might capture a glimpse of the full flowering of interspirituality.

1. *There is One God, the Eternal, the Only Being; none exists save God.*

2. *There is One Master, the Guiding Spirit of all Souls, Who constantly leads followers towards the light.*

3. *There is One Holy Book, the sacred manuscript of nature, the only scripture which can enlighten the reader.*

4. *There is One Religion, the unswerving progress in the right direction towards the ideal, which fulfils the life's purpose of every soul.*

5. *There is One Law, the law of reciprocity, which can be observed by a selfless conscience together with a sense of awakened justice.*

6. *There is One Brotherhood and Sisterhood, the human brotherhood and sisterhood, which unites the children of earth indiscriminately in the Parenthood of God.*

7. *There is One Moral, the love which springs forth from self-denial, and blooms in deeds of beneficence.*

8. *There is One Object of Praise, the beauty which uplifts the heart of its worshippers through all aspects from the*

seen to the unseen.

9.   *There is One Truth, the true knowledge of our being, within and without, which is the essence of all wisdom.*

10. *There is One Path, the annihilation of the false ego in the real, which raises the mortal to immortality, and in which resides all perfection.* [87]

If the sinking fortunes of religion are to be turned around, renewal will be necessary. A religious culture that is multimodal; traditional in procedure, but reformist in ethics; community-based but not tribally isolationist or antagonistic; and broadly interspiritual, will have the best chance to make such renewal possible. I'm skeptical that success is possible without plumbing our existing religious traditions, but there will be a place for the innovation of spiritual entrepreneurs like Bahá'u'lláh or Inayat Khan. (Although of course, that opens the door for hacks and swindlers, as well as sound new perspectives.)

Regardless, I find it hard to imagine that we have culturally transcended the need to be bound back to our community and our world, or ever will. We humans will never lose the urge to escape the chatter of our inner *I* and meld with the infiniteness of here, the timelessness of now, the source of love and compassion.

CHAPTER 10

SUCKING OUT THE POISON

After that last chapter, where I toot the horn for literally *every* other religion, and after having ignored the Buddha's wise counsel to Malunkya – not to chase after theoretical questions that don't really serve the path – I imagine the Buddha, were he were around to witness my bloviation, might be getting a bit exasperated with me right now. So this is the sheepish after-chapter where I try to remedy my standing with him.

The truth is, the Buddha was right – the truly important thing is not the Arrow, it is sucking out the poison. So I'll try to redeem myself in his eyes by offering a few thoughts on how to begin a Buddhist practice. If I can prod even one person who reads this to begin a life-changing habit, maybe my extravagant intellectual digression is not for nought.

Buddhism In Asia

First, we ought to openly acknowledge that the Buddhism written about in books like this one is not the Buddhism that is practiced by most of the people on this planet who identify as Buddhist.

How so?

More than the monotheistic religions, Buddhism in Asia is

a two-track religion. The monastic path contains study and meditation; that is the tradition we read about in books. The lay path might include those things, but more often it emphasizes rituals, holiday observances, and blends or co-exists with the devotional practices of the animist traditions.

I once met a Buddhist nun, American-born but working in Thailand, whose mission was to care for terminally ill villagers in the countryside. When trying to bring solace to their last remaining weeks of life, she knew better than to ask them to meditate; that practice never stuck with them. Instead, she would give them massages. Like meditation, massage was a way to bring them into their here and now. She would also perform rituals that sought to expunge bad karma; the local monks charged for this ritual, often more than the locals could afford. She gave them this deliverance for free. For these villagers, Buddhist practice was primarily devotional and ritualistic. There was little meditation, and little familiarity with Buddhism's foundational doctrines.

There is another key difference between Buddhism as practiced in Asia and Buddhism as we've talked about it here. I have emphasized the naturalistic aspects of Zen Buddhism because that is part of what attracts me about it. The Buddha was a concrete and practical thinker; the poisoned arrow metaphor he used when counseling Malunkya is a proof point for this. He cared about human outcomes – not dogmas or theorizing.

As the tradition has come down to me through American Zen, it comes with very little, if any, dogma attached to it. I can embrace the Buddha's teaching without abandoning my modern Western skepticism. But you don't need to dip very deep into Buddhist scripture and teachings to find a great deal that is supernatural.

For example, in the Pali canon there are many descriptions of how Mara, the personification of Death and Desire, comes to derail the Buddha's attempts to attain his great awakening. The Buddha commits to sitting beneath the bodhi tree and he resolves not to get up until he has attained enlightenment. In response, Mara sends his daughters to try to seduce him away from his endeavor with carnal desire. When that doesn't work, he sends successive armies of demons representing diverse varieties of weakness and distraction. Finally, Mara tries to trick him with his own double-talk. All his attempts to derail the Buddha's resolve fail, as the Buddha touches the earth beneath him and invokes it as witness and support. With the coming of the dawn the Buddha attains nirvana. Some might enjoy this as a poetic metaphorical depiction, but perhaps there are others who read this and mutter: "Demons... really?"

There are many, for example, who will be skeptical of reincarnation, a doctrine closely associated with many forms of Buddhism. Ask a Tibetan Buddhist, and you will get very detailed explanations as to how it works: how the karma you gain during your current life will determine your place in the next life. You don't need to be a hardened skeptic to find it a bit of a stretch. The idea of reincarnation is concordant with the Hindu concept of *atman*, or soul, but it is more difficult to reconcile with the Buddhist concept of *anatta*, or no-self. (Although, of course, there are rivers of written words trying to do just that.)

Some scholars suggest that the Buddha may have deliberately refrained from challenging the doctrine of reincarnation in order to avoid offending prevailing Hindu sensibilities and traditions.[88] Maybe so, maybe not – it is impossible to determine with certainty the mind of a man who lived two and a half millennia ago, and whose words were only committed to writing centuries after his death.*

---

* We know, however, that the oral tradition that maintained his teachings during the interlapse was rigorous and intentional.

I mention this not to provoke unproductive argument, but to acknowledge that the experience and beliefs I am sharing are just a small corner of the Buddhist world. Even if we consider only the strands of Buddhism that have made it to the West, it is still a broad cloth: the Tibetan tradition has received some celebrity affiliation, and has had a wise and charismatic figurehead in the Dalai Lama; the Theravada tradition has been carried on by organizations such as the Insight Meditation Society; and there has also been an impulse for a secular Buddhism, expressed by books such as *Buddhism Without Beliefs* and organizations such as Dharma Punx.

Beyond that, there are other Eastern religions like Taoism and the Hindu tradition that hold many elements in common with Buddhism. (Enkyo O'Hara Roshi, my teacher from the Village Zendo, has said she believes that Zen might have as much in common with Taoism as the Buddhist legacy.) There are also ostensibly secular – and sometimes commercialized – practices that incorporate meditation, like yoga and transcendental meditation. All these are paths up the same mountain.

Likewise, I have written in Chapter 7 of how the deist traditions use devotion to a God to quiet the Atomist Self and approach a Connected Self. If that cultural software has been installed in you and it is working, there is no reason to discard it. (Although as I elaborate below, I don't believe mindfulness and a belief in a deity necessarily exclude each other.)

All that said, I'll now describe one particular path up the mountain – because it happens to be the one I know.

Teacher And Community

My first exhortation is to not make the same mistake I made in my twenties. Do not become a Book Buddhist. Look, I see that book in your hand – it happens to be mine. Be warned, it is not close to sufficient. Add all its many, many siblings, and they will collectively still not be sufficient.

In the introduction to this book, I described a spiritual crisis in my forties – it was not my first. Earlier, in my late twenties, I wandered into that same dark forest. This first episode was even more painful than the latter one, but in retrospect, it was a very pure and special time as well. While normally I would wake up in the morning and wonder what I would have for breakfast, during this time questions like "who am I?" and "what am I?" were all I could think about from the moment my eyes opened. The search for answers consumed all my energy. I devoured spiritual reading: both classics like the Tao Te Ching, the Bible, and the Upanishads, as well as modern writers like Alan Watts and Huston Smith. I was especially taken by *Zen Mind, Beginner's Mind*, the collection of Shunryu Suzuki's talks, and many wonderful anthologies of Zen tales and sayings. Naturally, I started a meditation practice that followed the instructions I read in books. I bought an airplane ticket and travelled the European youth hostel circuit – the journey as metaphor. In Barcelona's Plaza Catalunya, I turned the last page of J.D. Salinger's *Franny and Zooey* – a novel about a woman in a spiritual frenzy, and her brother who tries to talk her down – and I was profoundly grateful to receive the perfect book at the perfect time in my life for it. (I might be less impressed by it today.)

What I didn't do was find a teacher and a community. It's not that I didn't learn from my spiritual crisis – it would be impossible not to! – but eventually the immediacy of that time passed. Prosaic questions like "what are you going to do with the rest of your life?" suddenly started to take primacy. I grieve sometimes for what my life might have been like had I found the right teacher then, instead of a decade and a half later.

Books can entertain you and inform you and inspire you, but it's very difficult to advance in the practice without a community and a teacher. According to one strand of Zen lore, there are only two people who achieved enlightenment without a teacher: the Gautama Buddha and Bodhidharma, the sage that brought Buddhism to China. (More on that fraught term "enlightenment" later.)

It is not that a teacher will be plying you with wisdom, lessons, and corrections constantly. In fact, they might say very little. But there is an immense hidden value to simply having a living example of an ethic and a way of being before you. (I suspect this is an underrated part of teaching in any sphere.) When you are in doubt about something, your teacher will be able to say "don't worry, it's normal" or "that's not quite right" and that gentle reassurance or redirection will be invaluable to you.

Likewise, the community you train with is very important. When you meet someone struggling with the same problems and asking the same questions as you, it can be reaffirming and encouraging. When you've known people for a while and you see their personality flower because of practice, it is eye-opening and can drive your own practice forward. Also, there will be people there who have been at practice for many more years than you, and the very fact that they are not formally teachers sometimes gives what they pass on to you an immediacy and a lack of self-consciousness – they don't even know they're teaching you.

Even meditation itself can change in the presence of others. There's a subtle but powerful difference between sitting alone and sitting in a group. In Zen, we say that we sit not just for ourselves, but for the whole world. That becomes clearer when you sit with others: you want to be still, not only for your own sake, but so you don't disturb the person beside you. And after a while in meditation, the soft, quiet focus of the group permeates the room and seems to seep into you through your pores. They become a part of you.

Zazen And Kimhin

Zazen is the name we give to meditation in Zen. Za means sitting – Zen means mind.

Zazen is not the entirety of Zen Buddhist practice; it also includes activities like ceremonies and rituals, talks and interviews, service and activism, the ethical code of the precepts, and more. Zazen, however, is the foundation of Zen.

Traditionally, Zazen is done sitting on a round cushion, called a zafu, that is placed over a square mat, called a zabuton. In contrast with many other traditions, in Zen we usually sit facing a blank wall, to avoid distractions.

A number of cross-legged positions are recommended for Zazen: full lotus, half lotus, and Burmese. Seiza, or kneeling pose, can also provide a solid foundation for sitting and is most often used with a bench, with the calves slipped underneath the seat, or sitting on a cushion resting on its side that is held fixed by the ankles. For those that have physical difficulty with sitting on the ground, sitting on a chair is also an option. If you do so, have both feet on the ground and try to avoid leaning against the back of the chair.

When deciding what position is best for you, try to find something that feels comfortable enough that you can sustain it for a length of time. However, also keep in mind that with time, your tendons and ligaments can sometimes become conditioned to the poses you take, especially if you are younger. The more ambitious lotus poses have a way of fixing your pelvis in a steady position, which in turn helps keeps your spine straight and still.

LOTUS

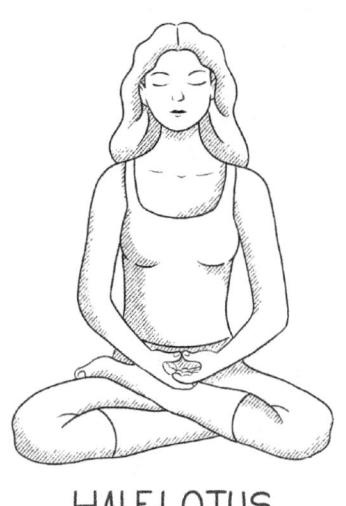

HALF LOTUS

BURMESE

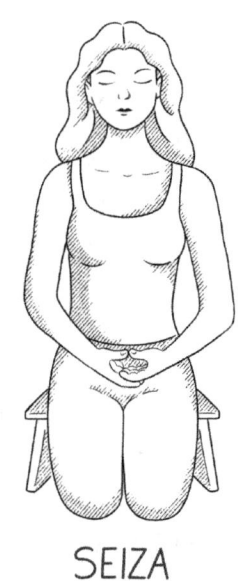

SEIZA

Once you have taken a seated position that feels solid and well-supported, lean forward from your waist and stretch your chest forward, arcing your back. Roll your hips forward and let your torso come back upright until you feel its weight centered over your seat. Now tilt from left to right and left again from your hips, searching for that sweet point where all your weight is distributed evenly over your center of gravity. Repeat this motion, with the tilts becoming smaller and smaller, until you find that point of balance.

Now imagine a string gently pulling upward from the crown of your head, elongating your spine. Let your tongue rest against the roof of your mouth, just behind your front teeth. Soften your gaze, allowing your eyelids to lower slightly, and direct your eyes down at a 45-degree angle toward the wall or floor. Your body should feel relaxed but with an upward energy. Let your shoulders rest comfortably on your rib cage. Don't try to force erectness; aim for a seat that feels comfortable and unstrained.

Take your left-hand fingers and place them over your right-hand fingers, and rest your hands on your lap. Let your thumbs touch together at their tips, so that your hand makes an oval. In the Indian tradition, this is called the Cosmic Mudra. It is said that if your thumbs sag, that's because you are going to sleep in your meditation. If your thumbs are pointing up, it's because your mind is too active.

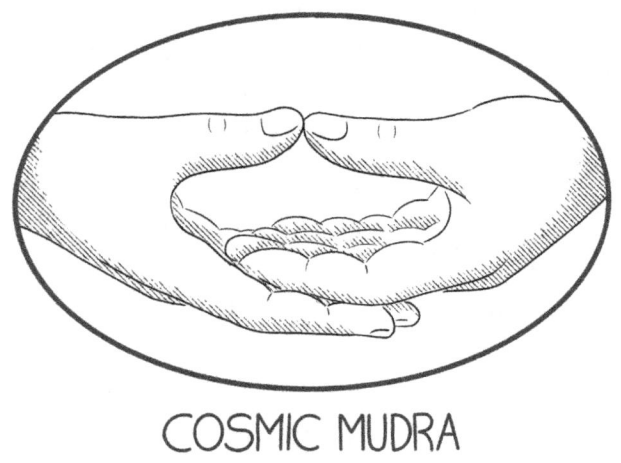

COSMIC MUDRA

Beginners are usually instructed to count their breath:

In, out, one. *In, out,* two. *In, out,* three.

Continue like this until you get to ten, then start with one again. If you become distracted by other thoughts, go back to one. Do not try to control your breathing but create space for it to deepen.

Once practitioners have become accustomed to counting breaths, they are instructed to let go of the counting and just pay attention to the breath. Notice the sensations in your belly, your lungs, your nose, your throat, as the air goes in and out.

Eventually, students are introduced to *shikantaza* – a Japanese term often translated as "just sitting." This is the practice of noticing whatever comes up in your awareness: physical sensations, sounds, thoughts, daydreams. When thoughts come up, you don't need to chase them away. Just let them come, neither rejecting them nor clinging to them.

While this is the common teaching sequence in the tradition

I know, it's a mistake to think of shikantaza as more "advanced" than counting breaths or following breaths. They are just different ways to meditate, and very experienced practitioners will sometimes count or follow their breath. Also, in some lineages of Soto Zen, shikantaza is the only sort of instruction given – it is what beginners begin with and what masters continue to practice, decades later.

What Zazen does is trick the DMN see-saw that I wrote about in Chapter 3. The act of observing our self is enough to tilt the see-saw away from its usual day-dreaming activities, if we commit sufficiently to observing. However, observation is a lightweight enough activity that the other end of the see-saw is not held down. When you practice Zazen for many hours a day while on retreat, your mind can get very quiet. It's like the see-saw stops in the middle, both sides floating in balance – you are neither in self-chatter, nor in a goal-oriented mode.

After finishing a sitting session at a Zendo, practitioners typically do a short session of *kimhin*, or walking meditation. This helps restore circulation to your legs without breaking you out of a meditative mindset. Usually you take very short steps, each step timed to the in-out of your breath. This break also serves to allow people to enter or exit the zendo.

Your practice need not end after you've left the Zendo. One of the beautiful (and challenging) aspects of Zen is that it doesn't need to stop. You can practice any time of the day by being present and aware of your body, your environment, and your companions.

Regularity And Intensity

In Japan, Zen has developed more as a monastic tradition than a lay tradition. Today, Zen communities are still trying to figure out (sometimes fitfully) how to adapt the life and practices of monks to the modern lifestyles of lay practitioners.

One important balance to strike is between regularity and intensity of meditation. It's essential to practice regularly, and we should set the goal to meditate at least once a day, even if it is for just a short period. But there is also a great deal to be gained from an intense period of practice. For that reason, Zen sanghas often offer *zazenkais*, day-long sitting periods, as well as *sesshins*, retreats of several days where the community practices intensely as you might in a monastery.

Breakthroughs in practice often come during sesshins because they allow you to immerse yourself into a deeper sort of meditation that usually isn't accessible in everyday life. An ideal lay practice should combine both regularity and intense periods of practice.

Combining With Other Practices

As mentioned before, Zen Buddhism takes no position on the existence of a deity. One of my teachers, the now departed Robert Kaku Gunn, was a UCC Pastor before he became a Zen Buddhist teacher at New Rochelle's Empty Hand Zendo. He never abandoned his belief in the God of the Bible. Likewise, you can find observant Jews and Muslims in many Zen sanghas. Not only is there no proscription of belief on the Zen side, I know of no proscriptions from the monotheistic religions of the sort of activities Zen Buddhists engage in.

Since it can require long hours of sitting that stiffen the body, many Zen practitioners supplement their meditation with physical disciplines. I've practiced yoga for over 20 years and find it an excellent preparation and complement to Zazen. Others I know have embraced practices like Qigong, Tai Chi, the Alexander Technique, and other body disciplines.

Rocky Shoals

There are some common pitfalls to avoid in any sort of mindfulness practice. The first two usually get in the way of your developing the habit. The third might happen if you love practice too much.

First: please do not translate states of mind into dualist notions of good and bad, virtue and sin, skill and incompetence. It might be tempting to think of distraction as bad, and a quiet mind as good when you meditate, but there's a problem with that: our minds are not prone to obedience. If I asked you to not think of Pink Elephants, it would probably be difficult for you to think of anything else. Likewise, if you play the role of a punitive taskmaster over your mind while you try to meditate, your practice is far less likely to prosper than if you are gentle with yourself. When a distraction comes in meditation, don't scare it off, don't chide yourself. Just observe the distraction and let it go on its own.

Indeed, it's probably a mistake to define success as a quiet mind. If you have set some time aside for yourself, if you have sat down to observe your mind at work, that in itself is a great success. Put your store in intentions rather than outcomes. Do not judge yourself by saying "I can't meditate. I just get too distracted!" Even the great masters become distracted in meditation sometimes.

The second obstacle is one that can't be circumnavigated – you just need to go through it. Many people feel pain and discomfort when they meditate. I don't mean physical pain, (we'll get to that shortly), I mean psychic pain. For some, mental activity is a sort of morphine. When it stops, the demons come out: fear, anger, self-loathing, sadness. There are edge cases where the demons come at every session, but at some point the demons come for nearly all of us. There's no getting around them. Sometimes, it takes real courage to sit. However, if you are strong enough to face yourself, you have a chance to move beyond the demons.

A word is needed about physical pain, too. Meditating for many hours a day can be physically taxing and is guaranteed to feel uncomfortable at times. If the pain comes in your legs, it's often due to restricted blood flow. A good quality zabuton is helpful there. Also, what some people do during long sesshins is use two different sitting positions, alternating between them to give areas of their legs a chance to rest and recuperate. If the pain problem you have is back pain, try getting ready for sesshin by doing exercises to strengthen your core. If the pain persists, some tutelage from a postural expert, perhaps an Alexander teacher, might help you minimize discomfort during your meditation. Finally, it should be said, most sesshins are going to include some sort of body practice to help work out the stiffness – take advantage of those opportunities to move and stretch.

The last shoal to avoid comes if you love the practice too much. For a small minority, meditation can become a way of hiding. Practice should be a way of opening you up to be more present and engaged in your everyday life, but for some it can become a way to avoid relationship struggles, social interactions, or professional challenges. It's okay if sitting starts feeling like a balm in your life, but if you are using it to avoid addressing issues and solving problems, then you are misusing the cushion.

Neither should you use Zazen as an emotional anesthetic. If you come to the cushion in a place of emotional upheaval, don't try to turn it off. Meet your emotions head on, where they are. They are something that is happening to your organism; like everything else that you experience when you are sitting on the cushion, just observe them without judgement.

The E-Word

I promised to broach the thorny topic of enlightenment. Here goes.

Who is an enlightened being? It is someone who lives beyond dualisms, beyond categories and distinctions, in a realm of not knowing. You can see the problem, then, with such a person raising their hand and saying: "I am an enlightened being!" Or even "she is an enlightened being!" Pronouncements like these reinforce the very boundaries an enlightened being is supposed to transcend.

If you come to a Zen teacher and say "I want to be enlightened," she might say: "You are already enlightened." Or she might say that just by sitting on a cushion you are enlightened. Quite a few calories are burned avoiding pointed questions about the matter. In fact, one of the most famous and most difficult koans[*] (and traditionally the first one to be assigned to students) is Mu:[89]

A monk asked Master Joshu: Does a dog have buddha nature, or not?

Joshu said 'Mu!'

The monk is asking a trick question. Of course, a dog has buddha nature! Even cat people know that. The monk certainly knows this. But for Master Joshu to say that a dog has buddha nature is to know – to make a distinction. (If you think this paragraph "solves" the koan you are mistaken.)

The following story shows another way to handle these inquiries.

---

[*] A koan is a sort of riddle or problem that is posed to the Zen student as a way to test and increase their understanding. Logic and reasoning will never solve a koan.

A student asked a master: "Do trees and streams have buddha nature?"

"Yes, trees and streams have buddha nature," the master replied.

"Do dogs and cats have buddha nature?" asked the student.

"Yes, dogs and cats have buddha nature."

"Do I have buddha nature?"

"No, you do not have buddha nature," pronounced the master.

The student objects: "Master, how can it be that animals, plants, and even inanimate objects have buddha nature, but I do not have buddha nature?"

"The animals, plants, and inanimate objects are not asking me this question."

That's all fine and good for probably a fictional sage, but at this stage you might be growing annoyed with me. You might protest that you did not come to this book to avoid objective inquiry – it was quite sciencey up until this damn chapter!

I'll try not to be evasive. I don't *personally* know anyone who has reported a Saul-falling-off-the-horse-to-become-Paul moment in their Buddhist practice. That is, I don't know of anyone who has crossed some threshold of practice where they suddenly attained the highest form of understanding, and certainly I have not experienced anything like that. This is not to say that such stories do not exist, nor is it to deny their possibility and authenticity.

However, my suspicion is that for the great mass of practitioners, the Way is a path with many signposts but no grand finish line. It is a practice of gradual accretion, of forward movement with occasional setbacks – a slow and steady development of the personality without grand fireworks or magical thresholds.

That is my take, but let me bring to your attention an old saying from the Taoist tradition:[90]

Those who understand, do not speak.
Those who speak, do not understand.

Has the ring of truth to it, but don't take my word for it. I'm speaking.

AFTERWORD

THE ROLL OF THE HAN

The Han is a large wooden block, usually suspended by rope from a ceiling or beam in a Zendo. It usually has a concave area where many years of impacts by a wooden stick have left their mark. When a session of meditation is due to start, the *jikido*, or timekeeper, knocks a distinctive rhythmic pattern on the Han to call practitioners to their cushions. Once everyone is seated, the jikido regulates the start and end of each sit with rings of the bowl bell.

Once all the sessions of the day are completed, the jikido returns to the Han and plays a pattern of rolls. Three rolls in total are heard, each roll being a series of knocks with the pauses between each hit becoming briefer and briefer, and the force of each impact becoming softer and softer. After the final roll, the jikido gives the Han an authoritative thwack, takes a couple of steps forward, and with hands in *gassho* recites the Evening Gatha:[91]

> *Let me respectfully remind you:*
> *Life and death are of supreme importance.*
> *Time swiftly passes and opportunity is lost.*

Each of us should strive to awaken.

Awaken!

Take heed. Do not squander your life.

This verse sounded very different to me when I first started practicing than it does now. It seemed to have an aggressive male energy that offended me a bit: domineering, tyrannical, presumptuous. Now I hear it quite differently. Of course, it still sounds like an exhortation to practice, but the way I think about how I can avoid squandering my life changes every time I hear the Evening Gatha. Sometimes I will even hear it as an exhortation to more hedonistic indulgence! Because an unenjoyed life is surely squandered, isn't it?

Maybe I have persuaded you that our species developed a number of faculties over the millennia, each with their own independent utility, and that these abilities were combined to create the majestic power of storytelling. That this new ability became not just a tool of communication and organization, but an internal mechanism of self-regulation and social connection: that we became Story Animals. Our minds turned into hives where the narrative often seemed more real than our actual experience. This mode was powerful, but it also left us discontent and vulnerable. And also: that there is a different way of being that can make us steadier, stronger, more connected and more content.

Or perhaps I have failed to convince you of any of this.

It doesn't matter. You're done listening to me, now I ask that you listen to you.

We Buddhists call our practice The Way, but really there are many Ways, aren't there? You have a Way. I've urged you to question your Self, but that is *not* the same thing as distrusting your organism. You probably know very well what you need to do to grasp opportunity in your life. Deep down, you know.

It might mean starting a practice like I have described... or taking more naps instead. It might be drinking less, or giving up social media. It might mean finding a new job, or making a better go at the one you've got. It might be asking a girl or boy out, or recommitting to a relationship. It might mean being more patient with your child, or giving your Mom a call. It might mean more work, or less work. It might mean bringing more people into your life, or protecting your moments of solitude. It might mean delivering food to the hungry and homeless, or a wild Friday night. Whatever it is, do it.

Take heed. Do not squander your life.

NOTES

Introduction: The Story Animal

[1] Andrew Whiten, "The Burgeoning Reach of Animal Culture," *Science* 372, no. 6537 (2021).

[2] This story appears in "Cūḷamālukyasutta," *Middle Discourses 63*. It and other Zen and Buddhist stories in this book have been paraphrased. The sources have usually been original, and sometimes I have referenced multiple translations.

[3] Brett D. M. Jones et al., "Magnitude of the Placebo Response Across Treatment Modalities Used for Treatment-Resistant Depression in Adults," *JAMA Network Open 4*, no. 9 (2021).

[4] Kuyken, Willem et al., "Effectiveness and Cost-Effectiveness of Mindfulness-Based Cognitive Therapy Compared With Maintenance Antidepressant Treatment in the Prevention of Depressive Relapse or Recurrence (PREVENT): A Randomised Controlled Trial," *Lancet* 386, no. 9988 (2015): 63–73.

Chapter 1: Something from Nothing

[5] Stanley L. Miller, "A Production of Amino Acids Under Possible Primitive Earth Conditions," *Science* 117 (1953): 528–529.

[6] Wikipedia, s.v. "abiogenesis," https://en.wikipedia.org/wiki/Abiogenesis. Provides an overview of the theories I've mentioned, and others.

[7] William Martin et al.,"Physiology, Phylogeny, and LUCA," *Microbial Cell* 3 (December 5, 2016).

[8] I draw biographical details for Lynn Margulis from a number of sources, including: *Lynn Margulis: The Life and Legacy of a Scientific Rebel*, ed. Dorion Sagan (White River Junction, VT: Chelsea Green, 2012); Bruce Weber, "Lynn Margulis, Evolution Theorist, Dies at 73," *New York Times*, November 24, 2011; n.a., "Lynn Margulis," *The Telegraph*, December 13, 2011.

[9] James A. Lake, "Lynn Margulis (1938–2011)," *Nature* (2011): 480, 458, https://doi.org/ 10.1038/480458a.

[10] n.a., *The Telegraph*, December 13, 2011.

[11] Quoted in Lynn Margulis, "Gaia Is a Tough Bitch," in *The Third Culture: Beyond the Scientific Revolution*, ed. John Brockman (New York: Simon and Schuster, 1996), https://www.edge.org/conversation/lynn_margulis-chapter-7-gaia-is-a-tough-bitch.

[12] Lynn Margulis, *Symbiotic Planet*, Science Masters Series (New York: Basic Books), 1.

[13] Margulis, *Symbiotic Planet*, 118.

[14] Margulis, *Symbiotic Planet*, 39.

[15] Lynn Margulis, Andrew Maniotis, James MacAllister, John Scythes, Oystein Brorson, John Hall, Wolfgang E. Krumbein, and Micahel J. Chapman, "Spirochete Round Bodies Syphilis, Lyme Disease and AIDS: Resurgence of 'the Great Imitator'?" *Symbiosis* 47 (2009): 51–58.

[16] Sagan, ed., *Lynn Margulis*.

[17] Ed Yong, *I Contain Multitudes: The Microbes Within Us and a Grander View of Life* (New York: Ecco, 2016). This is a joyous read on the topic of microbes.

Chapter 2: Parliament of Mind

[18] Biographical details about Darwin come from Hanne Strager's *A Modest Genius: The Story of Darwin's Life and How His Ideas Changed Everything* (Columbia, MD: CreateSpace, 2016).

[19] Pascale Tremblay and Simona M. Brambati, "A Historical Perspective on the Neurobiology of Speech and Language: From the 19th Century to the Present," *Frontier Psycholology* 15 (2024), doi: 10.3389/fpsyg. 2024.1420133. See also, Nasser Mohammed et al., "Louis Victor Leborgne ('Tan')," *World Neurosurgery* 114 (2018): 121–125.

[20] Society of Automotive Engineers, Aubertin, *Considérations sur les localisations cérébrales et en particulier sur le siège de la faculté du language articulé* (Paris: Masson, 1863).

[21] Chun Siong Soon, Marcel Brass, Hans-Jochen Heinze, and John-Dylan Haynes. "Unconscious Determinants of Free Decisions in the Human Brain," *Nature Neuroscience* 11 (2008): 543–545.

[22] Michael S. Gazzaniga, *The Social Brain: Discovering the Networks of the Mind* (New York: Basic Books, 1986).

[23] James Kingsland, *Siddhartha's Brain: Unlocking the Ancient Science of Enlightenment* (Boston: Mariner Books, 2017), 234.

[24] U Pu, in *Milindapañha* [Questions of King Milinda], Book 2, U Pu, trans. (New Dehli, India: Gyan Publishing House, 2024).

Chapter 3: The Atomist Self

[25] Gordon G. Gallup Jr., "Chimpanzees: Self-Recognition" *Science* 167 (1970): 86–87.

[26] L. David Mech, "Alpha Status, Dominance, and Division of Labor in Wolf Packs," *Canadian Journal of Zoology* 77, no. 8 (1999): 1196–1203.

[27] Marcus E. Raichle, Ann Mary MacLeod, Abraham Z. Snyder, William. J. Powers, Debra A. Gusnard, and Gordon L. Shulman, "A Default Mode of Brain Function," *Proceedings of the National Academy of Sciences* 98, no. 2 (2001): 676–682.

[28] Wei Luo, Biao Liu, Ying Tang, Jingwen Huang, and Ji Wu, "Rest to Promote Learning: A Brain Default Mode Network Perspective," *Behavioral Science* (Basel) 14, no. 4 (April 2024): 349.

[29] The categorization presented here is my own, though the underlying human needs have been widely studied and documented. They are sometimes organized differently – for example, in David Rock's SCARF model, which identifies five domains: Status, Certainty, Autonomy, Relatedness, and Fairness. David Rock, "SCARF: A Brain-Based Model for Collaborating With and Influencing Others," *NeuroLeadership* 1, no. 1 (2008): 1–9.

[30] Franz de Waal, "Two Monkeys Were Paid Unequally: Excerpt from Frans de Waal's TED Talk," YouTube, https://www.youtube.com/watch?v=meiU6TxysCg.

[31] Dimitrios Zagkas, Flora Bacopoulou, Dimitrios Vlachakis, George P. Chrousos, and Christina Darvir, "How Does Meditation Affect the Default Mode Network: A Systematic Review," *Advanced Experimental Medicine and Biology* 1425 (2023): 229–245.

[32] Caio Fábio Schlechta Portella, Ricardo Ghelman, Veronica Abdala, Mariana Cabral Schveitzer, and Rui Ferreira, "Meditation: Evidence Map of Systematic Reviews," *Frontiers in Public Health* 9 (2021), https://doi.org/10.3389/fpubh.2021.742715.

[33] Rachel Nuwer, "The World's Happiest Man Is a Tibetan Monk," *Smithsonian* (November 1, 2012).

[33] Thomas Merton, *The Way of Chuang Tzu* (New York: New Directions, 2010).

[34] Daniel Lim, Paul Condon, and David DeSteno, "Mindfulness and Compassion: An Examination of Mechanism and Scalability," *Plos One* 10, no. 2 (February 2015), https://doi.org/10.1371/ journal.pone.0118221.

Chapter 4: Carrots and Sticks

[36] Antonio R. Damasio, *Descartes' Error* (New York: Penguin Kindle Edition, 2005), 43.

[37]. Daniel Kahneman and Amos Tversky, "Prospect Theory: An Analysis of Decision under Risk," *Econometrica* 47, no. 2 (1979): 263–91.

[38] Philip Brickman, Dan Coates, and Ronnie Janoff-Bulman, "Lottery Winners and Accident Victims: Is Happiness Relative?" *Journal of Personality and Social Psychology* 36, no. 8 (1978): 917–927.

[39] Elizabeth Dunn, Timothy Wilson, and Daniel Gilbert, "Location, Location, Location: The Misprediction of Satisfaction in Housing Lotteries," *Personality & Social Psychology Bulletin* 29, no. 11 (2003): 1421–1432.

[40] World Health Org, "Depressive Disorder (depression)," *WHO*, https://www.who.int/news-room/fact-sheets/detail/depression.

[41] National Institute of Mental Health, "Suicide," https://www.nimh.nih.gov/health/statistics/ suicide.

[42] Dennis M. Bramble and Daniel E. Lieberman, "Endurance Running and the Evolution of Homo," *Nature* 432 (2004): 345–352.

[43] Ian Gilligan, *Climate, Clothing, and Agriculture in Prehistory: Linking Evidence, Causes, and Effects* (Cambridge: Cambridge University Press, 2010).

Chapter 5: Remembrance

[44] Ilias Tagkopoulos, Yir-Chung Liu, and Saeed Tavazoie, "Predictive Behavior Within Microbial Genetic Networks," *Science* 320, no. 5881 (2008):1313–1317.

[45] Robin I. M. Dunbar, "The Social Brain Hypothesis," *Evolutionary Anthropology* 6, no. 5 (1998): 178–190.

[46] The biographical details of Henry Molaison's life come from three sources: Philip J. Hilts, *Memory's Ghost: The Nature of Memory and The Strange Tale of Mr. M.* (New York: Touchstone, 1996); Suzanne Corkin, *Permanent Present Tense: The Unforgettable Life of the Amnesic Patient, H.M.* (New York: Basic Books, 2013); and Luke Dittrich, *Patient H. M.: A Story of Memory, Madness, and Family Secrets* (New York: Random House, 2016).

[47] Here is a friendly, thorough, and professional appreciation of Scoville's career: Andy Y. Wang, Diang Liu, Joseph N. Tingen, Harleen Saini, Vaishnavi Sharma, Alexandra Flores, and Ron I. Riesenburger, "The Life and Legacy of William Beecher Scoville," *Journal of Neurosurgery* 137, no. 3 (December 2021): 886–893.

[48] William B. Scoville, "Late Results of Orbital Undercutting: Report of 76 Patients Undergoing Quantitative Selective Lobotomies," *American Journal of Psychiatry* 117 (1960): 525–532.

[49] Dittrich, Luke. *Patient H.M.*

[50] Corkin, *Permanent Present Tense* (Basic Books, 2013). Many years ago, I

sent a much briefer blog-post version of this family history to Suzanne Corkin. The renowned memory researcher took the trouble to call me up, thank me for sending it, and compliment me on the writing. By a strange coincidence, Ms. Corkin grew up across the street from William Scoville in Hartford, and she was a close friend of a daughter of Dr. Scoville. She said she would forward the piece to her friend. Ms. Corkin said she was working on a book about H. M.; I promised her I would read it, and I did. That is the book this quote came from.

[51] Corkin, *Permanent Present Tense* (Basic Books, 2013).

[52] I learned this from a telephone conversation with Luke Dittrich, a grandson of Scoville who wrote a fine book about Henry Molaison, previously cited.

Chapter 6: The World Within

[53] Vicki Bentley-Condit and E. O. Smith, "Animal Tool Use: Current Definitions and an Updated Comprehensive Catalog," *Behaviour* 147, no. 2 (2010): 185–221.

[54] Thomas Suddendorf, *The Gap: The Science of What Separates Us from Other Animals* (New York: Basic Books, 2013), Kindle Edition, 149.

[55] Variations of the predictive process model have been proposed by Karl Friston, Anil Seth, Andy Clark, Jakob Hohwy, and others.

[56] Robert Wright, *Why Buddhism is True: The Science and Philosophy of Meditation and Enlightenment* (New York: Simon & Schuster, 2017). Kindle Edition, 193-194.

[57] This saying is most often attributed to the ninth-century Chinese Buddhist monk, Linji Yixuan.

[58] Andrew B. Newberg, Eugene G. D'Aquili, and Vince Rause, *Why God Won't Go Away: Brain Science and the Biology of Belief* (New York: Random House, 2002), Kindle Edition, 4–5.

[59] Newberg et al., *Why God Won't Go Away*, 6.

[60] Newberg et al., *Why God Won't Go Away*, 122–123.

[61] Daniel C. Dennett, *Breaking the Spell: Religion as a Natural Phenomenon* (New York: Penguin, 2007), Kindle Edition, 67

[62] Charles Darwin, *The Descent of Man, and Selection in Relation to Sex* (London: John Murray, 1871).

[63] Shanshan Li, Meir J. Stampfer, David R. Williams, and Tyler J. VanderWeele, "Association of Religious Service Attendance With Mortality Among Women," *JAMA Internal Medicine* 176, no. 6 (2016): 777–785.

[64] Harold G. Koenig, "Religion, Spirituality, and Health: The Research and Clinical Implications," *International Scholarly Research Network Psychiatry* 2012 (December 2012).

Chapter 7: Gods, Found and Lost

[65] Simon Baron-Cohen, Alan Leslie, and Uta Frith, "Does the Autistic Child Have a 'Theory of Mind'?" *Cognition* 21, no. 1 (1985): 37–46.

[66] Christopher Krupenye et al., "Great Apes Anticipate That Other Individuals Will Act According to False Beliefs," *Science* 354 (2016): 110–114.

[67] Matthew D. Lieberman, *Social: Why Our Brains Are Wired to Connect* (New York: Crown/Archetype, 2013), Kindle Edition, 106.

[68] Nashville Film Institute, "Kuleshov Effect: Everything You Need to Know," https://www.nfi.edu/kuleshov-effect/

[69] Catherine Caldwell-Harris, Caitlin Fox Murphy, Tessa Velazquez, and Patrick McNamara, "Religious Belief Systems of Persons with High Functioning Autism," *Proceedings of the Annual Meeting of the Cognitive Science Society* 33 (2011).

[70] David Brooks, "I Found Faith in a Crowded Subway Car," *New York Times*, https://www.nytimes.com/2024/12/24/opinion/david-brooks-journey-faith.html

Chapter 8: Symbols in the Machine

[71] Wikipedia, s.v. "Origin of Language," https://en.wikipedia.org/wiki/Origin_of_language

[72] Luisa Cacciante, Grazia Pregnolato, Stefano Salvalaggio, Stefano Federico, Paolo Kiper, Nico Smania, and Antonio Turolla, "Language and Gesture Neural Correlates: A Meta-Analysis of Functional Magnetic Resonance Imaging Studies," *International Journal of Language and Communication Disorders* 59, no. 3 (May/June 2024): 902–912.

[73] In a 2023 *New York Times* article, Sonia Shah tells the story of how cognitive scientist Tecumseh Fitch and evolutionary biologist Marc Hauser sent Chomsky a copy of their upcoming paper, which they viewed as a rebuttal of Chomsky arguments. They were surprised by an email from Chomsky requesting that they include him as a co-author on their next paper, which is this one: Marc D. Hauser, Noam Chomsky, W. Tecumseh Fitch, "The Faculty of Language: What Is It, Who Has It, and How Did It Evolve?" *Science* 298 (2002): 1569–1579.

[74] Most of my explanation of how Large-Language Models work is drawn from an LLM: ChatGPT.

[75] Heard on the April 12, 2024 *Ezra Klein Show* podcast, produced by the *New York Times*.

[76] Brian Christian, *The Alignment Problem: Machine Learning and Human Values* (New York: W. W. Norton, 2020), Kindle Edition, 144.

[77] Giovanni Di Pellegrino, Luca Fadiga, Luca Fogassi, Vito Gallese, and Giovanni Rizzolatti, "Understanding Motor Events: A Neurophysiological Study," *Experimental Brain Research* 91, (October 1992): 176–180.

[78] Leonardo Fogassi, Pier Francesco Ferrari, Benno Gesierich, Stefano Rozzi, Fabian Chersi, and Giacomo Rizzolatti, "Parietal Lobe: From Action Organization to Intention Understanding," *Science* 308, no. 5722 (2005): 662–667.

[79] Vilayanur Subramanian Ramachandran, *The Tell-Tale Brain: A Neuroscientist's Quest for What Makes Us Human* (New York: W. W. Norton, 2011), Kindle Edition, 23.

[80] Katherine Hawkes, John O'Connell, Neil Blurton Jones, Helena Alvarez, and Edward L. Charnov, "Grandmothering, Menopause, and the Evolution of Human Life Histories," *Proceedings of the National Academy of Sciences* 95, no. 3 (1998): 1336–1339.

[81] Greg J. Stephens, Lauren J. Silbert, and Uri Hasson, "Speaker–listener Neural Coupling Underlies Successful Communication," *Proceedings of the National Academy of Sciences* 107, no. 32 (2010): 14425–14430.

[82] The myth of the god Indra is told in many sources, but most famously in the *Avatamsaka Sutra* (also known as the *Flower Garland Sutra*).

Chapter 9: The Birth of Myth

[83] Paul Pettitt, *The Palaeolithic Origins of Human Burial* (New York: Routledge, 2011).

[84] Wikipedia, s.v. "History of Writing," https://en.wikipedia.org/wiki/History_of_writing

[85] I draw biographical details about Bahá'u'lláh from several sources, including https://www.bahai.org/bahaullah/life-bahaullah; https://en.wikipedia.org/wiki/Bahaullah; and the wonderful Filip Holm video on the Bahai faith: https://www.youtube.com/watch?v=75jZKEY2aRo

[86] I drew biographical details about Inayat Khan from several sources, including Wikipedia s.v. "Inayat Khan," https://en.wikipedia.org/wiki/Inayat_Khan; Hazrat Inayat Khan, https://inayatiyya.org/hazrat-inayat-khan/; and also from Filip Holm's video, Inayat Khan and Universal Sufism, https://www.youtube.com/watch?v=lal6rf5HhiY.

[87] Inayatiyya, "Ten Sufi Thoughts, https://inayatiyya.org/teachings/ten-sufi-thoughts/.

Chapter 10: Sucking Out the Poison

[88] For example, Johannes Bronkhorst, *Greater Magadha: Studies in the Culture of Early India* (New York: Brill Academic, 2007).

[89] This is the first koan in the collection entitled *The Gateless Gate*, compiled by Wumen Huikai, in 1228.

[90] This phrase is from the great Taoist masterpiece, the *Tao Te Ching*. Stephen Mitchell's translation is excellent, but perusing more than one translation at once has its rewards.

Afterword: The Roll of the Han

[91] This gatha appears in chanting books from both the Rinzai and Soto lineages of Zen. Its exact origin is unknown.

ACKNOWLEDGMENTS

You may recall that I dedicated this book to my teachers, so naturally, they belong here in the acknowledgments as well. First on that list are those who have generously transmitted the Zen tradition to me at the Village Zendo in New York, most notably Roshi Pat Enkyo O'Hara and Roshi Sinclair Shinryu Thompson. However, that sangha's great strength is that it has generated a great number of astonishing teachers – I owe thanks to many of them, but I will limit myself to just one more name: Roshi Ryotan Eiger taught me both at the Village Zendo and at the Empty Hand Zen Center in New Rochelle – I take his no-frills, no-fuss view of Zen to every cushion I alight upon.

Here in Buenos Aires, I am a member of the Vientos del Sur sangha, led remotely by Roshi Daniel Terragno from his home base in California. Zen, by the way, goes well with maté.

Of course, I have benefited from many teachers in topics other than Zen. I am especially grateful to those who taught me the craft of writing in its many forms. Building an architecture of meaning – word by word and sentence by sentence – is thrilling to me.

There were many readers who offered valuable suggestions and encouragement. Among them, David Parkes, Miriam Klevan, Fritz Oltmann, and Nate Jebb. One reader deserves special mention: Jordan Scott. His shrewd perspective on a host of topics – pre-history and wolves among them – corrected many of my misperceptions and undoubtedly enriched this text.

I was fortunate to find the perfect developmental editor for this project. James Kingsland has previously served as Science Editor for *The Guardian*. He is also the author of *Siddhartha's Brain*, a superb exploration of the neuroscience of meditative practice. As a practicing Buddhist himself, he was sympathetic to my viewpoints, as well as amply qualified to comment on the scientific dimensions of this work.

Finally, I was lucky enough to find a meticulous, sharp-eyed, and opinionated copyeditor under the same roof as me: my wife, Jennifer Griffith. She has been a source of unwavering support and encouragement, and she is among the teachers I dedicate this book to. From her, I receive the lessons I most need to learn.

ABOUT THE AUTHOR

William Gadea is a Peruvian-American writer, filmmaker, and Zen practitioner with a passion for storytelling in all its forms.

He was raised in Peru, Australia, and the Dominican Republic before coming to the United States to study film at New York University. After film school, he was drawn to the theatre, where his plays have been produced regionally and off-off-Broadway, as well as published and anthologized.

While pursuing playwriting, he supported himself with day jobs on Wall Street and in network news. Tired of having two jobs, he decided to take night courses in animation, and was soon helping to create children's programming for outlets such as MTV, Nickelodeon, and PBS.

In 2012, he founded IdeaRocket, an international animation studio that has created videos for more than 30 Fortune 500 companies, as well as a host of innovative startups. It was through his work for pharmaceutical and healthcare companies that he developed a knack for explaining complex scientific concepts with metaphor and narrative.

Lately, he has been creating videos for his YouTube channel, *What Is Mind?* He lives in Buenos Aires, Argentina with his wife Jennifer, and their cat Hildy.